一般集成论

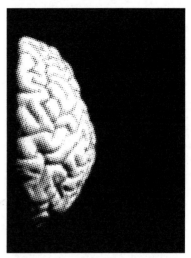

唐孝威 著　王大辉 译　王小潞 校

浙江教育出版社·杭州

图书在版编目（CIP）数据

一般集成论 / 唐孝威著；王大辉译 . -- 杭州：浙江教育出版社，2023.12
ISBN 978-7-5722-6835-9

Ⅰ．①一… Ⅱ．①唐… ②王… Ⅲ．①脑科学—研究 Ⅳ．① Q983

中国国家版本馆 CIP 数据核字（2023）第 212552 号

一般集成论
General Integration Theory

唐孝威 著　王大辉 译　王小潞 校

责任编辑	王荟捷　江　雷		责任校对	朱晨杭
美术编辑	韩　波		责任印务	沈久凌
装帧设计	融象工作室 _ 顾页			

出版发行　浙江教育出版社（杭州市天目山路 40 号）
图文制作　杭州林智广告有限公司
印刷装订　杭州捷派印务有限公司
开　　本　880mm × 1230mm　1/32
印　　张　15.25
插　　页　6
字　　数　400 000
版　　次　2023 年 12 月第 1 版
印　　次　2023 年 12 月第 1 次印刷
标准书号　ISBN 978-7-5722-6835-9
定　　价　79.00 元

如发现印装质量问题，影响阅读，请与我社市场营销部联系调换。联系电话：0571-88909719

前言

脑是自然界最复杂的系统，脑的活动是自然界最复杂的物质运动形式。当代科学技术的一个重大课题是研究和了解脑的工作原理，以及学习和应用脑的工作原理。本书在讨论向脑学习的基础上，提出了一般集成论的理论。

考察脑的结构和功能时，可以看到脑具有许多不同的层次，包括分子、基因、突触、神经细胞、神经回路、功能专一性脑区、脑功能系统、整体的脑，以及心智与行为等。在微观和介观水平上，神经系统存在各种不同的集成统一体，如突触、神经细胞、神经回路等；而在宏观水平上则有脑功能系统和整体的脑等集成统一体。

脑内不同层次有不同种类的集成成分，基于它们之间的各种相互作用，构成不同形式和多种功能的集成体。脑的不同层次上存在许多类型和多种形式的集成作用和集成过程。脑内不同层次的集成体进一步集成为统一的、具有复杂结构和复杂功能的整体的脑，涌现出丰富多彩的心智，并且产生多种多样的行为。脑内的集成是随时间发展的动态过程，这种集成过程是通过脑内的集成作用以及脑与环境的集成作用实现的。

脑为我们提供了一个研究集成作用和集成过程的"实验室"。本书根据脑的实验事实,从脑的结构与功能集成、脑的信息集成、脑的心理集成等方面,讨论脑内许多类型和多种形式的集成作用和集成过程的特性。我们提出,需要建立一门研究脑的集成现象的特性和规律及其应用的新的学科,可以把它命名为脑集成论。这门学科着重在脑的系统水平上对脑的集成现象的特性和规律及其应用进行研究。

在考察神经系统的集成作用和集成过程后,我们还提出需要建立一门研究神经系统的集成现象的特性和规律及其应用的新的学科,可以把它命名为神经集成论。这门学科涉及神经系统所有层次的集成现象,比较侧重在微观和介观水平上对神经系统的集成现象的特性和规律及其应用进行研究。

当我们把视野从脑转向自然界、科学技术领域和人类社会时,可以注意到这样的事实:集成作用和集成过程不仅在脑的活动中起重要的作用,而且在自然界、科学技术领域和人类社会中广泛存在。在自然界、科学技术领域和人类社会中,不同层次和不同种类的集成成分,基于它们之间的各种相互作用,集成为不同层次、不同形式的集成统一体,并且在一定条件下涌现出新的特性。

自然界、科学技术领域和人类社会中不同领域的集成作用和集成过程分别具有各自特殊的规律,但它们又具有共性,可以用一些一般性的概念来对各种不同的集成作用和集成过程的共性进行统一的描述。

本书在向脑学习以及研究不同领域中相关实验事实的基础上，归纳各种集成作用和集成过程的一些一般性的概念，如优化、全局化、互补、协调、符合、同步、绑定、涌现、适应、同化、集大成、大统一等，可以用它们描述各种集成现象的共性。

虽然人们早已有集成的观念，也有过对几种具体的集成现象的一些分散的讨论，但是还没有对自然界、科学技术领域和人类社会中的集成现象，包括其中的集成作用和集成过程，进行过统一的、系统的研究。基于自然界、科学技术领域和人类社会中广泛存在各种集成现象的事实，本书提出需要建立一个新的学科，这个学科要研究不同领域中的集成现象，包括其中的集成作用和集成过程的一般性规律及其实际应用。我们把这个学科命名为一般集成论。

本书包括三篇。第一篇讨论脑的集成现象和脑的集成理论，介绍脑的一些实验事实，考察脑的集成作用和集成过程，并提出建立神经集成论、脑集成论和仿脑学等学科。

第二篇讨论一般集成论。我们根据不同领域中广泛存在各种集成作用和集成过程的事实，归纳集成现象的一般性概念，提出一般集成论的论点，再说明一般集成论与前人的系统论等理论的联系和区别。

第三篇讨论一般集成论的应用。举出一般集成论在多种领域中应用的例子。例如，在生物学和医学中的应用、在心理和行为科学中的应用、在知识创新和学科建构中的应用、在工程技术科学中的应用、在文化教育领域中的应用等。

本书作者在神经科学和认知科学领域中是初学者，对工程技术科学和人文社会科学领域知识的了解亦不足，书中可能存在错误和缺点。请各方面专家给予宽容，并且不吝指教，也请广大读者对本书的不足之处提出批评和建议。

本书是浙江省科学技术厅资助完成的研究成果。

目录

第一篇 脑的集成现象和脑的集成理论 001

第一章 集成的脑 003
1.1 脑的结构和功能 004
1.2 脑复杂网络和脑内通信 010
1.3 脑的活动和能量消耗 016

第二章 脑的集成 022
2.1 脑的结构与功能集成 023
2.2 脑的信息集成 026
2.3 脑的心理集成 032

第三章 神经集成论、脑集成论和仿脑学 041
3.1 神经集成论 041
3.2 脑集成论 043
3.3 仿脑学 046

第二篇 一般集成论 049

第四章 不同领域的集成现象 051
4.1 自然界的集成现象 051
4.2 技术领域的集成现象 054
4.3 学科交叉的集成现象 056
4.4 人类社会的集成现象 058

第五章 探索一般集成论 061
5.1 一般集成论理论 061
5.2 全局和模块 069
5.3 还原和综合 071
5.4 绑定和联合 072
5.5 重建和优化 074
5.6 临界和涌现 075
5.7 互补和协调 077
5.8 符合和同步 079
5.9 适应和同化 082
5.10 集大成和大统一 083

第六章 一般集成论的特点 085
6.1 在向脑学习基础上发展的一般集成论 085
6.2 一般集成论与专门集成论 088
6.3 一般集成论与系统论等理论的联系和区别 092

第三篇 一般集成论的应用 099

第七章 生物集成论 101
7.1 生物集成与生物进化 102

7.2 活细胞的集成　104
7.3 人体的集成　108

第八章　心理集成论　115
8.1 心理相互作用与心理集成　115
8.2 意识的集成　121
8.3 认知的集成　128
8.4 心智的集成　132
8.5 心智与行为的集成　139

第九章　知识集成论　142
9.1 知识集成与学科交叉　143
9.2 选择性注意的集成模型　149
9.3 心理学学科体系的集成　152
9.4 认知科学理论的集成　157

第十章　工程集成论　165
10.1 大科学计划的集成　166
10.2 大型实验装置的集成　167
10.3 医学影像技术的集成　171

第十一章　教育集成论　176
11.1 智能的集成　176
11.2 教育内容的集成　184

附录　187

参考文献　209

名词简释　220

被災記念日本赤十字社事業発展

第一編

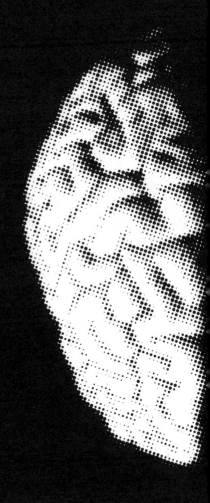

哪者且各县著复交的关系，哪
的动总是各月从各事物质里出化的状
式，根据脑内的大重客版重者，我们可
以了解脑内的事故迎象，并且研究脑
的事故迎律。

这一周以脑的客观事实为主，说
明脑各事故的一体，并且其一要
做眼内在的关系的形式的事故作用和
事故关系，在此基础上，我们追出
立一门讲述脑的事故迎象和迎律的专
科——脑事故，但的容按出运立种
事故迎和的脑各则事等标。

这一周的第一章讲述事故的脑，
第二章讲述的事故，第三章讲述用由
练事故法，哪事故法和作练法。

第一章　集成的脑

我们首先来考察自然界最复杂的系统——人脑。

脑具有层次性结构：从生物大分子、基因、亚细胞结构、神经细胞（神经元与神经胶质细胞）、神经元簇、神经回路到功能专一性脑区、脑功能系统和脑的整体。脑内这些不同层次的结构和它们的功能是脑整体活动的基础。对于脑这个复杂系统，需要从各个层次来研究其结构、功能和工作原理。

脑的复杂性不仅表现在脑的层次的复杂性、脑的结构的复杂性、脑的功能的复杂性、脑内网络的复杂性、脑内通信的复杂性、脑和环境相互作用的复杂性，而且表现在丰富多彩的脑的高级功能——心智现象的复杂性以及人的行为的复杂性。

这一章讨论集成的脑，从脑的结构和功能、脑内复杂的网络和脑内通信、脑的活动和能量消耗等方面介绍脑的一些实验事实，说明脑是集成的统一体。

1.1 脑的结构和功能

脑的层次可以大致划分为微观、介观和宏观三个不同的水平。从小的方面到大的方面来看，微观水平包括分子、亚细胞结构和神经细胞；介观水平包括神经元簇和神经回路；宏观水平包括功能专一性脑区、脑功能系统和脑的整体。这里所说的微观和物理学中微观的含义不同，是指分子和细胞。

脑内这些不同水平的神经结构具有不同的空间尺度。脑的微观水平、介观水平和宏观水平神经结构的空间尺度相差很大。例如，神经细胞的细胞体的数量级大致是 10^{-6} 米，功能专一性脑区的数量级大致是 10^{-2} 米。脑内这些不同水平的神经结构的活动过程具有不同的时间尺度。脑的微观水平、介观水平和宏观水平的神经活动过程的时间尺度相差很大，例如神经元电脉冲的数量级大致是 10^{-3} 秒，脑功能系统活动的数量级大致是 10^{-1} 秒。

我们在实验中可以用多种技术研究脑的这些不同水平的结构和功能。例如，分子神经生物学技术、神经遗传学技术、神经解剖学技术、神经电生理技术、显微成像技术、脑结构成像技术、脑功能成像技术等。

脑功能成像技术用于脑的宏观水平的功能研究，能够对脑的功能活动进行无创伤的、动态的成像，如功能磁共振成像（fMRI, functional Magnetic Resonance Imaging）、单光子发射计算机断层成像（SPECT, Single Photon Emission Computerized Tomography）、正电子发射计算机断层成像（PET,

Positron Emission Computerized Tomography)、脑电成像(EEG, Electroencephalography)、事件相关电位(ERP, Event-related Potentials)、脑磁成像(MEG, Magnetoencephalography)等。

神经科学家对脑的结构和功能进行了广泛的实验研究，不断取得新的进展，因此目前对脑有了相当多的了解(Purves et al., 1997; Gazzaniga et al., 1998; Kandel et al., 2000; 陈宜张，2008)。但是由于脑非常复杂，至今仍有许多问题有待了解。

从微观水平来看，人脑大约有10^{11}个具有多自由度的神经元和数量更多的神经胶质细胞，神经元之间形成约10^{14}—10^{15}个突触联结。神经元有细胞核、细胞质、细胞膜和细胞骨架。神经元的细胞膜是镶嵌蛋白质的脂双层，膜上有许多由大分子蛋白质构成的复合体，如受体及离子通道。离子通道是细胞内外离子穿越细胞膜的通道。

神经元内部含有各种神经递质和神经调质。神经递质是在神经信号传递中起作用的化学物质，它们的种类很多，如乙酰胆碱、L-谷氨酸等。神经调质是调制神经细胞生化反应的化学物质，它们的种类也很多，如多巴胺、去甲肾上腺素等。

神经元具有多种多样的形态。它们有细胞体和许多分支结构：有大量的小分支即树突，神经元通过树突接收神经信号；通常有一根细长的纤维即轴突，神经元通过轴突传出神经信号。轴突具有传导神经信号的功能，树突在神经信息处理中也有重要作用。张香桐(1997)很早就研究过树突的生理功能。

突触是一个神经元和另一个神经元相互作用的部位。神经元通过大量的突触和其他神经元相互作用，一个神经元平均形成10^3—10^4个突触联结，因此神经元之间构成非常复杂的神经网络。突触部位具有动态性和可塑性，在神经信息处理中起重要的作用。

脑内神经胶质细胞的数量大约是神经元数量的10倍，它们不仅对神经元起支持和营养作用，而且对神经元的活动起调节作用，影响神经元的功能。此外，神经胶质细胞释放的分子调节脑血管的收缩与舒张，使局部脑血流量的供应与神经活动的水平相适应（陈宜张，2008）。

脑的介观水平的神经结构和功能介于微观水平和宏观水平之间，是由微观水平过渡到宏观水平的桥梁。一些神经元和神经胶质细胞连接成介观的神经回路。神经回路的研究是当前神经科学研究的热点之一。实验上对神经回路的分子和细胞机制以及神经回路的功能和可塑性进行了大量研究，理论上也有许多神经回路的模型。有研究者专门对脑的介观水平的神经动力学进行了研究（弗利曼，2004；Liljenström & Århem，2008）。

从宏观水平来看，脑位于颅腔内，受到脑膜的保护，脑内有血管分布和血液循环，为脑正常的生理活动提供养料、清除废料。脑具有复杂的结构，由大脑、间脑、中脑、小脑、脑桥、延髓等部分组成；中脑、脑桥和延髓合称为脑干。脑和脊髓是神经系统的中枢部分，它们和周边神经构成了身体内完整的神经系统。

大脑有左、右两个半球，它们之间有纤维连接。大脑表面凹凸不平，有许多皱褶。主要的沟回把大脑分为额叶、顶叶、枕叶、颞叶以及脑岛。大脑由灰质和白质组成。大脑皮质是神经细胞集中处，形成了灰质；大脑皮质下方是由神经纤维束组成的白质。脑细胞活动所需的氧和葡萄糖由脑内血管通过血流供应。

大脑皮质包括许多具有不同功能的专一性脑区，如视觉皮质区、听觉皮质区、躯体感觉皮质区、运动皮质区、语言皮质区等。这些功能专一性脑区组成脑的功能系统，如接受、加工和储存信息的脑功能系统等；脑功能系统再组成整体的脑。

在脑功能系统方面，我们曾根据实验事实提出了脑的四个功能系统学说。《脑的四个功能系统学说》（唐孝威，黄秉宪，2003）一文介绍了鲁利亚（Luria）（1973）的脑的三个功能系统学说，并且阐述了在鲁利亚学说基础上发展的脑的四个功能系统学说。下面引用该文在这方面的一些说明：

> 鲁利亚对大量的脑损伤病人进行了临床观察和康复训练，观察到脑的一定部位的损伤会引起一定的心理功能障碍；但脑的一种功能并不仅仅和某一部位相联系，脑的各个部位之间还有紧密的联系。鲁利亚（1973）根据研究事实，把脑分成三个紧密联系的功能系统，并且提出了脑的三个功能系统学说。
>
> 鲁利亚在《神经心理学原理》一书中，阐明脑有以下三个功能系统：第一个系统是保证、调节紧张度和觉醒状态的功能系统，这个功能系统的相关脑区是脑干网状结构和边缘系统；第二个系统是接受、加工和储存信息的功能系统，这个功能系统的相关脑区是大脑皮质的枕叶、颞叶、顶叶等；

第三个系统是制定程序、调节和控制心理活动与行为的功能系统，这个功能系统的相关脑区是大脑皮质的额叶等。人的行为和心理活动是这三个功能系统协同活动的结果。脑的三个功能系统的学说对了解脑的整体功能具有重要意义。

然而我们注意到，除了这三个功能系统外，评估和情绪等心理活动对于脑的整体功能同样是必不可少的。由于当时实验资料的限制，在鲁利亚的学说中并没有包括与评估和情绪等心理活动有关的功能系统，而目前实验提供的大量事实则越来越表明这些心理活动的重要性。

从实验事实看，评估功能是在许多心理活动中普遍存在的（Edelman & Tononi，2000）。机体在进化过程中形成了适应个体和种系生存和发展要求的、对外界环境输入信息的意义进行评估的系统。在个体脑内先天的评估结构基础上，机体根据过去的经验和当前的需要形成评估的标准；评估系统将输入信息的意义与评估的标准进行比较，从而给出评估结果；个体由评估的结果对信息按重要程度决定取舍及处理，对可能作出的反应做出抉择；经评估和抉择做出的决定，通过调节、控制的功能系统对机体状态进行调控，并对外界环境做出反应（黄秉宪，2000）。

脑内信息处理过程的每一步都需要对信息进行评估，因此脑内评估是在心理活动中不断进行的。脑内评估系统具有可塑性，它的评估标准随着个体学习过程而形成和发展，并且不断发生变化。

脑内的评估-情绪功能系统与鲁利亚提出的第二、第三功能系统有类似的组织结构。它也是一个多层次的系统。

情绪系统是评估-情绪功能系统中比较基础的一部分，它对情境的整体信息进行评估，并产生强烈的主观体验和反应。在脑的高级部位，评估系统能够对特殊的信息，甚至对具体思维结果做精确的评估。

脑内对外界信息进行评估的结果，还会引起个体的情绪体验：符合个体需要或愿望的信息，有肯定性的评估结果，并可能产生正的情绪体验；不符合个体需要或愿望的信息，有否定性的评估结果，并可能产生负的情绪体验（Arnold，1960；Lazarus，1993）。

脑内杏仁核对奖惩相关的事件记忆起重要作用，所以杏仁核是与评估功能相关的脑区。中脑侧背盖区、黑质等处的多巴胺神经元能对预测的奖励与实际奖励的误差做出反应（Waelti et al.，2001），这也可能是评估系统的一部分。

边缘系统等与情绪功能有关的脑区（Le Doux，1996）也是评估—情绪功能系统的一部分。此外，前额叶的一部分可能是评估—情绪功能系统的高级部位。

为了弥补鲁利亚的脑的三个功能系统学说没有涉及评估-情绪功能的不足，我们在三个功能系统学说的基础上，把评估-情绪功能系统列为脑的第四个功能系统。因为对信息意义进行评估以及由此产生情绪体验，是脑的基本功能，而前面提到的调节紧张度和觉醒状态的功能系统，接受、加工和储存信息的功能系统，以及编制程序和调节控制行为的功能系统，都没有包括评估和情绪的功能。评估-情绪功能系统有别于其他几个功能系统，所以有必要把它专门列为另一个功能系统。

在以上讨论的基础上，我们发展鲁利亚的脑的三个功能系统学说，提出脑的四个功能系统学说，认为脑内存在四个相对独立而又紧密联系的功能系统：第一功能系统——保证、调节紧张度和觉醒状态，第二功能系统——接受、加工和储存信息，第三功能系统——制定程序、调节控制心理活动和行为，第四功能系统——评估信息和产生情绪体验。这四个功能系统集成为整体的脑，人的各种行为和心理活动，都是这四个功能系统相互作用和协同活动的结果。

从以上的说明可以看到，复杂的脑是集成的脑。

1.2 脑复杂网络和脑内通信

神经系统的重要特性是神经信号的传导和处理。神经电生理学在脑的微观水平上，对神经信号传导和处理进行过深入的研究（Purves et al., 1997；顾凡及，梁培基，2007）。实验表明，一个神经元和同它有突触联结的另一个神经元的神经信号传递是通过电过程和化学过程的耦联而实现的。

神经元细胞膜上有对电压敏感的离子通道。神经元细胞内外液体中有各种离子，如钠离子、钾离子、氯离子等。在神经元处于静息状态时，细胞内外液体中离子成分和浓度不同，使细胞膜两侧存在电位差，细胞膜内的电位相对于细胞膜外为负电位，称为静息电位。

当神经元受到刺激而处于活动状态时，细胞膜外的钠离子选择性通透细胞膜，从膜外进入膜内，使细胞膜电位发生变化。这时产生的连续变化的电位称为分级电位。当膜电位的变化达到一定值时，产生持续时间短而波形和幅度不变的电脉冲，称为动作电位。这种由细胞膜的离子通透性变化造成的动作电位沿着神经元轴突单向传递。

一个神经元和同它有突触联结的另一个神经元的突触部位存在突触间隙。轴突末梢的细胞膜内有许多包含化学物质的囊泡，当动作电位脉冲传递到轴突末梢的突触前膜时，会使这些囊泡把其中的化学物质释放到突触间隙中，它们在突触间隙中扩散到达突触后另一神经元的细胞膜，与膜上特异性的受体结合，就会引起突触后神经元细胞膜的离子流动。

化学突触有不同的种类。有对突触后膜起兴奋性作用的兴奋性突触，还有对突触后膜起抑制性作用的抑制性突触。如果兴奋性突触后细胞膜的电位变化达到一定值，那么就会引发突触后神经元的动作电位。

动作电位的特点是"全或无"，也就是说，要么不产生动作电位（无），要么产生动作电位，其波形和幅度维持不变（全）。在信号传输过程中，动作电位脉冲幅度不会衰减。外界刺激可以使神经元中产生一连串动作电位，动作电位脉冲序列携带有关刺激特性的信息。

在脑的宏观水平上，脑内通信是在脑功能子系统集成的脑整体网络中进行的。我们曾在《脑功能原理》（唐孝威，2003a）一书中对此进行过讨论。下面就引用其中的几段文字进行说明：

> 神经解剖学和神经心理学为脑功能子系统提供了实验证明。人的躯体各部分的感受器分别投射到大脑感觉皮质的特定区域；而大脑运动皮质的特定区域则分别控制身体各部分的运动，因此感觉系统和运动系统分别有脑内精确定位的网络，以完成特定的功能。在高级功能方面，布罗卡（Broca）和韦尼克（Wernicke）的研究表明，语言有一系列可分开的、可定位的子系统（Kandel et al., 2000）。
>
> 有研究者指出，脑功能的分离与整合是一个基本的实验事实（Frackowiak, 1997）。一方面，脑由大量功能子系统组成，它们相对独立地进行信息加工（Zeki, 1993），某一类脑损伤可能影响到一类脑功能，但其他脑功能则不受影响（McCarthy & Warrington, 1990）。另一方面，脑功能子系统之间互相连接而形成整体的神经网络，脑的整体功能依赖于功能子系统之间的相互作用（Fuster, 1997）。利用脑功能成像技术获得了大量实验资料，这些资料进一步表明，脑内存在许多功能分离而又相互协作的脑功能子系统。

脑内大量神经回路组成了具有不同功能的、相对独立而互相连接的功能子系统。一个功能子系统包括了完成同一功能的有关脑区。在各个功能子系统之间有连接通路，这些连接通路是以大量神经通路为基础的等效道路。许多功能子系统以不同方式广泛连接而形成脑的整体网络（黄秉宪，2000）。[许多研究者用功能磁共振成像研究脑的自发血氧水平依赖性活动的网络模块组织（He et al., 2009）。]

一个功能子系统可以用复杂的电子线路来等效地描述，这个电子线路对输入信号进行各种变换、组合、分析和记录。这里讨论的输入、输出神经信号，都是指脑内功能子系统之间传递的信号，而不是外界的物理刺激。这种等效电子线路除线路本身外，还包括：输入信号的输入端和输出信号的输出端，实现调控作用的多种门控结构，与其他功能子系统的等效连接通路等。一个功能子系统有输入端和门控输入，可以接受两类输入信号：一类输入信号通向输入端，引起功能子系统脑区激活；另一类输入信号通向门控输入，使功能子系统脑区的活动受到调控，例如抑制或增强。

一个功能子系统总有多个输入端，在简单讨论时可以只考虑只有一个输入端有信号输入而其他输入端无信号输入的情况。一个功能子系统还具有多个输出端，在简单讨论时可以只考虑其中一个输出端的输出，输出信号还可以反馈。功能子系统的门控结构类似于电子线路中的与门、非门、或门等，起甄别、符合、反符合等作用。例如，在有抑制的门控作用时，功能子系统的输入信号受到抑制；在有符合的门控作用时，功能子系统的输入信号得到增强。一个功能子系统与其他功能子系统之间由等效连接通路广泛连接，这些连接通路是整个系统不可缺少的组成部分。

近年来，脑的默认网络（default mode network）引起人们广泛的关注。我们在《意识全局工作空间的扩展理论》（宋晓兰，唐孝威，2008）一文中曾介绍过脑的默认网络，这里引用该文几段文字进行说明：

> 近年来一个普遍得到证实的现象是，大脑的一部分区域，在静息或进行简单的被动感觉刺激任务时较进行主动的刺激-反应任务时更活跃（Shulman et al., 1997; Raichle et al., 2001; Raichle & Snyder, 2007）。在完成各种加工外界刺激信息的任务状态下，这些脑区通常表现为负激活，并且空间分布非常一致，且无论是在静息状态下还是在任务状态下，这些脑区的自发低频血氧水平依赖性信号都有较强的时间相关性，即它们之间以功能连接（functional connectivity）的方式组成网络（Raichle et al., 2001; Greicius et al., 2003; Fransson, 2005），被称为默认网络。其中扣带回后部/前楔叶，前额叶中部/扣带回前部腹侧是较为核心的两个区域。在静息状态下，这些脑区的能量代谢在全脑中最高（Raichle & Mintun, 2006），提示这部分脑区在静息时进行

着某些有组织的活动，而这种活动在外在任务时受到抑制，表现为负激活。

脑内存在的以功能连接方式组织起来的默认网络的活动模式，称为脑功能的默认活动模式（Raichle & Snyder, 2007）。脑功能的默认活动模式表明，脑是一个高度自组织化的器官，在没有环境刺激时也在进行有规律的活动。上述默认网络是比较特殊的一个，根据其独立于任务的活动水平下降的特点，认为默认网络内脑区在静息时进行着某种与外界刺激无关的活动，这些活动可能与监控内外环境、形成自我意识、情景记忆提取以及维持大脑清醒的意识觉知状态等过程有关（Raichle & Snyder, 2007）。

默认网络与一些和注意及工作记忆相关的脑区呈负相关关系（Fox et al., 2005），且认知负荷对默认网络的活动强度具有调节作用，随任务难度增加，其活动水平下降（McKiernan et al., 2003）。还有研究发现，默认网络的自发低频血氧水平依赖性信号与大脑后部广泛的感觉皮质，包括视觉、听觉、体感皮质活动呈负相关（Tian et al., 2007）。虽然这与福克斯（Fox）等人的结果有出入，但都提示了默认网络的活动与加工外部信息所需的大脑活动之间可能存在拮抗关系。

默认网络内脑区参与的过程相当广泛，根据与控制条件激活对比的结果，认为这些脑区可能涉及情景记忆（Greicius & Menon, 2004）、自我参照加工（Northoff & Bermpohl,

2004)、情绪加工（Maddock，1999）、主体感加工和第一人称视角加工（Vogeley et al., 2004; Gallagher & Frith, 2003）以及维持意识觉知（Raichle et al., 2001）等活动。由于静息时个体的认知加工的复杂性，没有一个"任务"的激活可以同时"复制"出这些区域，因而无法给出默认网络静息态功能的直接证据。但稳定的负激活网络说明了这些复杂认知过程中存在的共性。从目前的研究证据看，这些区域远离相对较低级的感觉、运动皮质，其活动总是会受到外在刺激-反应任务的抑制，推测它可能进行着与外部刺激无关的内部信号的加工。

我们从以上的说明可以看出，在脑复杂网络和脑内通信方面，复杂的脑是集成的脑。为了进一步理解脑的整体同步性质及其信息传输和集成作用，人们正在研究神经网络上的雪崩和脑功能连接组学（Beggs & Plenz, 2004; Biswal et al., 2010）。

1.3 脑的活动和能量消耗

脑的活动包括生理活动和心理活动。脑内有血液循环、氧代谢、葡萄糖代谢等过程，同时有脑内神经电活动和神经化学反应等，这些都是脑的生理活动。脑有认知、学习、记忆、语言等高级功能，这些都是脑的心理活动。脑的生理活动和心理活动是紧密联系和统一的。

在脑区活动方面,我们曾讨论过脑区的激活和相互作用(唐孝威,2003a),认为脑内功能子系统的一个基本特性是相关脑区的激活。一个功能子系统可以处于不同的状态,脑区静息的状态是功能子系统的基态,在脑活动时,功能子系统处于激活态。激活态有不同的激活程度。脑区激活是指有关脑区内部神经元的电活动以及与此相联系的生物化学反应。描述脑区激活的参量是激活水平,它是脑区激活程度的量度。激活水平反映相关脑区内部大量神经元活动的总和,包括神经元兴奋和抑制的总效应,也反映相关脑区内部生物化学反应及能量代谢的水平。在同一个功能子系统内部,脑区激活有空间分布。在脑的系统水平上,我们只考虑功能子系统总的激活水平,而不讨论它内部细致的激活分布。

功能子系统的另一个基本特性是与其他功能子系统之间的相互作用。功能子系统之间互相传递神经信号,包括神经元的电信号以及生物化学信号。在脑整体网络中,各个功能子系统间复杂的相互作用是通过功能子系统间连接通路传递信号来实现的。这些信号是指脑内功能子系统间的信号,而不是原初的外界物理刺激。外界物理刺激先转换成神经信号,再送到有关的功能子系统的输入端。描述信号的参量是信号强度,包括信号的幅度与持续时间。输入信号引起功能子系统脑区的激活,脑区激活后会通过连接通路传递信号而与连接脑区作用。描述连接通路的参量是通路效能。

如果用激活与相互作用的观点考察脑功能子系统与脑整体网络,那么我们似乎可以对感觉、知觉、动作、注意、学习、记忆、意识等各种脑功能有一个大致统一的理解。例如，感觉器官是脑内感觉功能子系统与外界环境之间相互作用的输入界面，外界物理刺激在感受器中转换为神经信号而传输到脑内感觉功能子系统；知觉是感知觉功能子系统在自下而上与自上而下的信号作用下的激活与相互作用；运动器官是脑内运动功能子系统与外界环境之间相互作用的输出界面，脑内一些功能子系统的信号输入运动功能子系统，其输出信号作用于运动器官，产生并控制运动；注意是脑内控制系统与受控制的功能子系统相互作用，使受注意的功能子系统的激活通过符合作用而增强，并抑制其他不受注意的子系统；学习是脑功能子系统多次激活导致网络连接通路的易化；记忆的贮存是脑功能子系统内相关结构的形成，记忆的提取是这些结构的重新激活；意识和无意识的脑活动都是脑区激活与相互作用，而意识的涌现相当于在调节、控制系统作用下有关的脑功能子系统激活态的相变。

从信息观点看，在脑功能活动中脑内进行着复杂的信息加工（黄秉宪，2000）。在脑的系统水平上，脑内信息的载体是脑功能子系统的脑区及其激活态。在脑内信息加工中，信息的获取、编码、贮存、提取、传递、变换、产生等过程都是通过脑区激活与相互作用实现的。

我们在《无意识活动与静息态脑能量消耗》（宋晓兰，陈飞燕，唐孝威，2007）一文中曾介绍过静息态脑的能量消耗的现象，下面引用该文几段文字进行说明：

近年来，脑功能成像技术的发展使我们可以研究各种不同条件下的脑能量消耗。研究发现，成人的脑质量占全身总质量的2%，但其能量消耗占全身总能量消耗的20%，是其质量比例估算值的10倍。而当环境中存在需个体努力集中注意的刺激即个体处在任务状态中时，脑增加的能量消耗却不到总能量消耗的1%（Raichle & Mintun，2006）。也就是说，大脑静息态能耗远远大于任务引起的能耗增加。

由此可见，脑在静息态下的"内禀"活动不仅表现为血氧水平依赖性信号的低频波动，还消耗着大量能量。脑不是只对环境刺激起反应的被动组织，而是时时刻刻进行着某种自组织的活动。有人指出，脑在静息态时的能耗80%以上被用于神经信号加工（Shulman et al.，2004），我们认为这些"内禀"活动的实质是脑内信息加工过程。

以上讨论的是个体清醒并觉知的状态。实验给出，在意识状态发生改变的情况下，脑的能量消耗会发生改变。如药物引起的麻醉会使个体全脑葡萄糖代谢大幅度下降，由不同药物引起的麻醉而导致的代谢下降幅度从31%到68%不等（Alkire et al.，1995；1997；1999）。另一项资料来自无觉知状态的植物人，一般认为植物人处于无意识状态（Laureys，2005）。研究发现植物人大脑能量代谢比控制组低30%—40%（Schiff et al.，2002）。从外界刺激引起的脑反应来看，大部分研究结果表明植物人对刺激的反应仅限于较低级的感觉运动皮质，但也有个案研究发现有些植物人的脑会对语言信息做出反应，从功能磁共振成像实验数据看，有些植物人可以根据实验者的要求做心理表象的活动（Naccache,2006）。另外还有来自深度睡眠（慢波睡眠）的资料表明，在此阶段，脑代谢活动总体下降；同时，正电子发射断层成像和行为实验表明，深度睡眠阶段大脑进行着与巩固白天记忆以及加工白天遗漏的信息有关的活动（Miller,2007）。

有人提出过静息态脑的能量消耗现象几种可能的解释（Raichle, 2006）。例如，这部分能量被用于静息状态下不受控制的自发的认知活动，如白日梦等。但是既然同属意识觉知部分，这部分能量消耗要超过任务状态下增加的能量消耗是说不通的。另一种解释是，这部分能量可能用于神经元之间兴奋性和抑制性联结的平衡，这可能是大脑自发的自组织活动的一个原因（Buzsáki, 2007），但没有确切的证据支持这种解释。第三种解释看起来较为合理，即这种内在固有的活动是在不断地进行着信息的保持以更好地应对环境的变化。尽管提出过这些解释，但是至今还不了解静息态脑能量消耗的本质，因此有人借用天文学中的名词，称它为脑的"暗能量"（Raichle, 2006）。

我们曾对静息态脑的能量消耗的机制提出一种解释（宋晓兰，陈飞燕，唐孝威，2007），认为静息态脑内大量能量的消耗，除了如赖希勒（Raichle）和明顿（Mintun）（2006）所说的用于神经细胞的修复、蛋白质运输等过程外，大部分用于无意识加工过程，其中主要是在没有外界任务情况下自发产生的无意识加工过程，它们包括了上面提到的为应对环境变化做准备的过程。这些加工过程并不随着外界任务的出现即注意资源的大量占用而停止。

我们指出，无意识活动的内容虽然不被觉知，但相关的脑区处于激发态，只是其激发的水平比较低，没有到达意识阈值。这些激发脑区的信息加工活动是消耗能量的，虽然单个脑区消耗的能量水平较低，但因为存在大量的、并行的、自发进行的无意识活动，就造成了持续不断的高能量消耗。

在这个意义上，在没有外界任务的情况下，大脑也并没有休息（Miall & Robertson, 2006），而是一直处于活动状态。相对于外界任务引起的瞬时"激活"而言，静息态的"基线"活动是持续不断的（Gusnard & Raichle, 2001），脑进行着的持续不断的信息加工，即我们指出的无意识加工。

从脑的活动和能量消耗来看，脑的工作是在脑内物质集成、能量集成、信息集成等集成过程中实现的，复杂的脑是集成的脑。

第二章 脑的集成

上一章介绍脑科学的基础知识,通过脑与心智的一些实验事实,说明脑是集成的脑。这一章则根据脑与心智的一些实验事实,说明脑的集成现象。

集成的脑和脑的集成是紧密联系着的。这一章和上一章讨论内容的不同之处在于:上一章用集成的观点说明脑是集成的统一体,这一章则用集成的观点说明脑内各种集成现象,特别是各种集成作用和集成过程。

脑内集成是一个动态过程,是脑内各部分集成为脑统一体的过程,这种过程是随时间发展的,是通过脑内相互作用以及脑与环境相互作用实现的。脑的各种相互作用不但将脑内各部分集成为统一的整体的脑,而且还由神经系统将身体各个生理系统和各种器官集成为统一的整体的身体。

谢林顿(Sherrington)对中枢神经系统的集成作用进行了开创性研究。他在《神经系统的整合作用》(Sherrington,1906)一书中详细讨论了脊髓反射作用,阐述了神经系统不同层次的整合作用。

他提出了"中枢神经系统的作用在于整合作用"的著名论断。"整合"就是本书所说的"集成"。

脑内有许多不同种类的集成作用和集成过程，下面几节分别考察脑的结构与功能集成、脑的信息集成和脑的心理集成。

2.1 脑的结构与功能集成

这一节从脑的结构与功能方面讨论脑的集成作用和集成过程。结构集成指脑结构方面的集成作用和集成过程，功能集成指脑功能方面的集成作用和集成过程。在生物学中，结构是功能的基础，要了解功能就得了解结构。脑的结构集成和功能集成有紧密的联系，结构集成为功能集成提供基础，而功能集成又促进结构集成。

脑的结构集成和功能集成是在不同的时间尺度上进行的，这些时间尺度有很大的跨度。从种系的进化过程来说，脑的集成过程的时间尺度是数十万年到数百万年以上。从个体一生发育和生长过程来说，脑的集成过程的时间尺度是数十年到百年以上。从个体生命的一定阶段的学习过程来说，脑的集成过程的时间尺度是数天到数年。从个体的某种心智活动过程来说，脑的集成过程的时间尺度是数十毫秒到数十秒。下面着重从脑的进化过程和脑的发育过程来说明脑的结构集成和功能集成。

经过长期进化过程中的自然选择，特别是由于劳动，人类不但发展了对生存有意义的神经结构，而且逐渐形成了复杂的脑功能系统。埃克尔斯（Eccles）在《脑的进化》（Eccles，2004）一书中对脑的进化进行了详细的阐述。大量事实表明，人脑的结构和功能是在长期进化中逐步发展的，进化过程是脑的集成过程，今天的人脑是长期集成作用和集成过程的产物。

复杂的脑是由数量巨大的神经元和神经胶质细胞构成的，它们是怎样集成为整体的脑的呢？普维斯（Purves）等人（1997）在《神经科学》一书中给出了关于脑的发育的许多资料，说明了从胚胎到成人的脑，脑的发育涉及神经系统的分化、脑的节段化、神经元的迁移、突触的形成、神经回路的构建等复杂的过程。这表明在脑的发育中存在各种集成现象，个体脑发育过程是脑的集成过程。

前面提到，脑是具有许多层次的复杂系统。在生物大分子和神经细胞到整体脑之间，存在许多个中间层次。由神经元和神经胶质细胞集成为不同的中间层次的结构，中间层次结构又集成为整体的脑。

脑具有可塑性。在脑的不同层次和不同方面，脑的可塑性有不同的表现，它们分别具有不同的时间特性和空间特性。下面分别讨论突触的可塑性、脑区的可塑性、脑内网络的可塑性，以及损伤脑区的可塑性。

突触具有可塑性。实验表明，当一个神经元的突触传递兴奋信号且突触后神经元同时发放时，这个突触的连接强度增加（Hebb，1949）。突触连接强度在神经活动中发生改变的特性称为突触可塑性，突触的可塑性是脑内网络可塑性的微观基础。

脑区具有可塑性。对一个脑区来说，每次激活都导致这个脑区以及它和其他脑区间连接的变化。例如，某种长时间的训练使大脑皮质中与训练相关的区域发生重组，在一定情况下，相关区域的皮质会增大。

脑复杂网络具有可塑性。脑区及其连接的可塑性使得脑复杂网络是可塑的。实验表明，实践活动对脑内网络的发展有重要影响。例如，训练过程使脑内网络发生变化。脑内网络的可塑性是学习和记忆的基础。

从脑的发育来说，在新生儿发育过程中，脑内神经元的轴突和树突快速增长，脑的重量不断增加。儿童期和青少年期的脑也在不断发展，脑内网络越来越复杂。波斯纳（Posner）和罗特巴特（Rothbart）(2006) 曾研究过儿童发育中脑内注意网络的发展过程。

感觉器官损伤的研究表明，如果某种外周感觉器官受到损伤，大脑皮质中与损伤器官对应的区域会发生重组，使它不再具有原来的功能，而是逐渐发展成感知或处理其他感觉信息的系统，这说明大脑皮质具有很强的可塑性。

损伤脑区也具有可塑性。如果大脑皮质中某种功能区域发生局部的损伤，那么通过康复之后，受损伤脑区的功能可以部分地由其他脑区代偿，也就是说，执行原来功能的系统可能会转移到其他脑区。

总之，脑在不断进行塑造，脑的塑造过程是脑的结构与功能的集成过程。

2.2 脑的信息集成

脑是处理信息的器官，脑内通信是脑的重要特性。脑的信息集成是脑内集成作用和集成过程的一个重要方面，脑内大量不同的信息集成为整体信息。这一节从脑的信息处理方面讨论脑的集成作用和集成过程。

脑的信息集成和上一节讨论的脑的结构集成和功能集成有密切的联系。脑的信息集成和脑的结构集成的关系是：脑的结构集成是脑的信息集成的基础。脑的结构是为实现脑内信息加工而构筑的。除脑内神经细胞组成神经回路外，脑内神经细胞之间可以有远程的连接，这些神经连接有精确的空间分布，为脑的信息集成提供了结构上的基础。

脑的信息集成和脑的功能集成的关系是：脑的信息集成是脑的功能集成的基础，脑的许多功能集成由脑的信息集成过程实现。谢林顿（1906）指出，神经系统的整合作用是通过神经信号的传导完成的。神经信号可以快速传导，而且神经传导具有精确的时间分布。

在单个神经元的水平上，一个神经元能够将输入的兴奋性信号和抑制性信号集成为统一的反应。下面引用坎德尔（Kandel）等人（2000）在《神经科学原理》一书中对神经信号集成的说明：

> 由单个突触前神经元产生的突触电位一般比较小，不足以使突触后神经细胞激发到形成动作电位的阈值。但大脑皮

质的每一个神经元不断接收来自其他神经元的许多输入信号，有些信号是兴奋性的，其他是抑制性的。有些信号强度较大，另一些信号强度较弱。这些输入信号与神经元树突的不同部位相接触，它们可以互相增强或抵消。

当突触后细胞接收兴奋性输入时，可能也会接收抑制性输入。抑制性输入会阻止动作电位的产生。任何一个兴奋性突触或抑制性突触的输入的总效应是由许多因素决定的，如突触的位置、大小和形状，以及其他突触的邻近程度和相对强度等。

在突触后神经元中，通过神经元集成的过程，将一些竞争性的输入进行集成。这种神经元集成是整体神经系统面临的决策任务在细胞水平的反映。也就是说，在任一时刻，一个神经细胞要对动作电位发放或不发放作出决定。谢林顿把神经系统由竞争性抉择作出决定的活动称为神经系统的集成作用。他把决策看作脑的最基本特性之一（Sherrington, 1906）。

脑内神经信息的编码问题，即神经信息在脑内如何编码、表达和加工的问题，是脑内信息集成的基本问题之一。我们在《神经元簇的层次性联合编码假设》（唐孝威等，2001）一文中曾经探讨过神经元簇的信息编码，下面引用该文的部分内容：

> 脑的信息编码研究由来已久，从1949年赫布（Hebb）

提出的经典细胞群假设（Hebb，1949），到1972年巴洛（Barlow）的单个神经元的编码假设（Barlow，1972），以及1996年藤井（Fujii）等人提出的动态神经元集群时空编码假设（Fujii et al.，1996），不同观点间的争论始终在进行。其中争论的一些重要问题是：是单个神经元还是神经元集群编码刺激信息？是神经元动作电位出现的明确时间还是脉冲的平均发放速率携带信息？由于神经系统的高度复杂性，利用现有的实验手段还不能彻底解决神经信息编码原理，目前已有的几种神经编码理论在解释神经系统工作原理方面都存在不同层次的困难。

神经生物学实验表明，神经系统处理信息具有几个明显特点。首先是多样性，即可以辨认同一目标的不同形态。例如，对人的不同表情或不同年龄的同一张人脸的模式识别，以及对其他的同属一类但形态、大小和颜色各异的物体的识别等。其次是鲁棒性，即个别神经元的死亡及损伤并不导致相关神经信息的丢失。最后是层次性，即对刺激信息的处理分特征层次和抽象层次。例如，祖母的各个细节与祖母这个抽象概念是两个不同层次。

巴洛单个神经元编码假设的基本点是平均发放速率编码假设、最优刺激概念、单个神经元等于单个功能及功能连续性假设。巴洛理论在高级认知功能方面的直接扩展就是祖母细胞假设。该编码理论的明显困难是组合爆炸，困难的根源在于其基本假设：需要存在代表客观事物多个特征及属性之间各种组合的编码细胞，但有限的脑细胞不足以满足这种需要。就像冯·德·马尔斯堡（von der Malsburg）（1981）曾指出的那样，该理论产生的困难比其能解决的问题还多。另外，单个神经元编码理论也无法解决鲁棒性问题，即大脑中认知细胞的产生和消亡带来的记忆管理问题。

赫布经典细胞群假设面临两个密切相关的困难：重叠困难和绑定问题。重叠困难指两个刺激同时到达时，由于该理论基于细胞群内的所有神经元平均发放速率的增加来识别一个编码群，故引起无法分辨这两个细胞群的问题。重叠困难的根本原因是赫布经典细胞群缺乏内部结构，而所要表达的外界知识是有层次和结构的。绑定问题是皮质内整合多个平行通路中的信息时出现的困难。例如，红色圆圈和绿色三角形同时呈现在视场时，脑中颜色区中代表红色和绿色的细胞群同时兴奋，而形状脑区中代表圆圈和三角形的细胞群的平均发放率亦同时提高，在赫布经典细胞群框架下，无法完成正确的绑定。

除以上两种主要编码理论外，还存在其他一些有一定影响的学说，如藤井为代表的图元假说，冯·德·马尔斯堡的同步振荡编码、位置编码理论、时间编码假说，以及近年来引起广泛关注的动态神经元集群时空编码理论等。需要指出的是，有确定的神经生物学实验证明，即使同一皮质区也存在完全不同的编码范式，所以我们不排除在同一脑区存在两种以上编码的可能性。

为了克服目前几种编码理论的困难并解释更多的实验事实，我们提出一个新的假设，它融合了祖母细胞假设和赫布经典细胞群编码假设的优点，称为神经元簇的层次性联合编码假设。我们认为神经编码的基本单元是神经元簇，一个神经元簇由一群功能和定位都比较接近的神经元构成；每个神经元簇有选择性的特征表达，对该特征刺激反应最强烈，而对其他刺激的反应随该刺激与最优刺激间的差距而减弱；参与编码的神经元簇具有等级性，存在编码抽象概念的神经元簇及编码具体属性的神经元簇；神经元簇编码的一个基本性质是联合表征，即不同层次的神经元簇可以被绑定在一起共同表达复杂刺激。

　　这个假设的建立除考虑了前述神经系统处理信息的几个特点外，还基于下列实验事实：（1）皮质存在功能柱结构，位于同一功能柱内的皮质神经元对某一特定的传入刺激有相似的放电反应；（2）嗅觉功能柱对某种气味分子存在最大反应，而且对相近的分子具有一定反应灵敏范围。其他功能柱也有类似的特性。但最优刺激响应神经元簇与非最优刺激响应神经元簇之间的关系，尚需深入研究。

　　神经元簇的概念显然与祖母细胞不同，因为基本单元是神经元簇，而不是单个神经元；它也与赫布经典细胞群含义不同，因为神经元簇有选择性特征表达，并有层次性联合编码，而赫布经典细胞群是一些以解剖学联结为基础，以相关发放为指导而组织起来的一个神经元集群。虽然在某些方面神经元簇类似赫布经典细胞群，但神经元簇的识别依据一定时间窗口（例如100ms）内部成员的共同活动即发放率的同时增加来完成，神经元簇内所有神经细胞的解剖位置和功能都非常相似。

神经元簇的层次性联合编码的优点是可以避免目前几种编码假设的困难。首先可以解决鲁棒性问题。由于每个神经元簇含有相当多个（例如10^3个或更多）神经元，个别神经元的损毁或死亡不会影响整个神经元簇对其特定刺激的表达。其次，神经元簇编码假设神经元簇具有内部结构，具有多个特征及属性的复杂事物可以激活代表各个不同特征和属性的一群神经元簇，它们的恰当绑定共同编码该复杂刺激。联合表征需要特定绑定，特定绑定的一个可能解决方案是更高级脑区的选择性注意机制。至于是否需要由同步振荡进行绑定，还要进一步研究。另外，这种联合表征可以在不同层次间进行，即编码抽象概念的神经元簇及编码具体属性的神经元簇可联合表达各种复杂事物。最后，在神经元簇编码假设中存在多个功能相似的神经元簇，可以解决多样性问题。例如，祖母不同表情的脸，可由不同的神经元簇编码和联合表征。

在脑的整体水平上，《脑功能原理》（唐孝威，2003a）一书曾讨论脑区激活和相互作用中神经信号的集成作用:(1) 未激活脑区在不接受输入信号时，保持其原来状态；激活脑区在不接受输入信号时，其激活水平随时间衰减;(2) 输入信号使脑区激活，脑区激活水平随输入信号强度的增大而提高;(3) 激活脑区输出信号到达连接脑区，信号强度随激活脑区激活水平的升高而增大，并随连接通路效能的提高而增大;(4) 激活脑区对连接脑区作用，使它们之间连接通路的效能提高，连接脑区又反作用于前面脑区。

此外，还要提到脑的集成作用和集成过程的另一个方面，即脑、身体与环境的集成。脑不是孤立存在的封闭系统，而是处于身体与环境之中的开放系统。环境包括个体所处的自然环境和社会环境。

脑、身体与环境不能分离，心-脑-身系统不断地与自然环境和社会环境相互作用，并在相互作用中不断地塑造脑与心智。在认知科学中，曾经有几种不同的观点研究心-脑系统和环境的作用，如具身的观点 (Lakoff & Johnson, 1999)、情境的观点 (Brooks, 1999)、动力系统的观点 (Thompson & Varela, 2001) 等。脑、身体与环境之间的各种相互作用促进了脑、身体与环境的集成。

2.3 脑的心理集成

心智活动是脑的高级功能。脑的心理集成是指脑内心智活动的集成作用和集成过程。我们在《心智的无意识活动》(唐孝威，2008a) 一书中提出，觉醒注意成分、认知成分、情感成分和意志成分以及这些成分之间的相互作用构成心智的整体。该书曾对心智活动有如下的说明：

　　心智有觉醒注意成分。一定的觉醒是心智活动的基础，个体觉醒才会有各种主观体验。觉醒可以处于不同的程度，反映个体心智的整体觉醒状况。个体的觉醒程度是随时间变动的。觉醒与心智的其他成分有关，如觉醒程度受情感影响，也与意向有关。

　　心智有认知成分。个体的主观体验有具体的内容。在认知过程中，个体知道自己觉知的是什么，在此基础上，还知道觉知内容所具有的意义。认知过程有信息加工，心智的内容是脑内加工的各种信息以及信息的意义，其中包括脑接收的内、外环境输入的信息和脑发出的支配动作的输出信息等。

　　在认知方面，感觉、知觉、记忆、注意、思维、语言等都属于心智活动。感觉是客观事物作用于感觉器官，而在脑中产生的对事物的个别属性的认识。知觉是客观事物在脑中产生的对事物整体的认识。记忆是脑对外界输入信息进行编码、存储和提取的过程。记忆是心智活动的重要方面，个体既有对当前信息进行加工的短时的工作记忆，还有长时存储的长时记忆。按存储信息的性质分，长时记忆有情景记忆和语义记忆。注意是心理活动对一定对象的指向和集中。思维是脑对信息进行分析、综合、比较、抽象和概括的过程。语言是人类用来交流思想的符号系统，语言过程是一种心智活动。

　　心智还有情感成分和意志成分。个体在情绪和情感方面的主观体验以及在意志方面的主观体验都具有心理学的意义。情绪和情感是人对客观事物的态度体验和相应的行为反应。意志是有意识地支配和调节行为，并且通过克服困难实现预定目标的心理过程；个体总是对自己的活动有意向。

自我意识包括自我认知、自我体验、自我控制等。自我认知是人对自己的洞察和理解。自我体验是伴随自我认知而产生的内心体验。自我控制是自我意识在行为上的表现。

从上面的一些说明中，我们可以看到心智的多样性和复杂性。心智的各种成分之间有紧密的联系，心理集成是通过这些成分之间的相互作用而实现的。上一章介绍了脑的四个功能系统，这里说明心智的各种成分和脑的功能系统的关系，以及它们之间的相互作用。以下文字引自《统一框架下的心理学与认知理论》（唐孝威，2007）一书：

> 脑的四个功能系统不是孤立、无关的，它们之间存在相互作用。它们各自的功能活动以及它们之间的相互作用和协调工作，形成了脑的整体活动。上面提到心智的四种主要成分，即觉醒注意成分、认知成分、情感成分和意志成分，这些心理活动成分和它们之间的相互作用组成整体的心理活动。心理活动四种主要成分和脑的四个功能系统之间有密切的关系。
>
> 脑的四个功能系统及其相互作用是心理活动成分以及它

们之间相互作用的物质基础。心理的觉醒注意成分主要基于脑的第一功能系统的活动。心理的认知成分主要基于脑的第二功能系统的活动。心理的意志成分主要基于脑的第三功能系统的活动。心理的情感成分主要基于脑的第四功能系统的活动。

心理活动各种成分之间的相互作用都分别有其脑机制。心理活动的觉醒注意成分和其他心理成分之间的相互作用，主要是通过脑的第一功能系统和脑的其他几个功能系统之间的相互作用来实现的。脑的第一功能系统保证、调节紧张度和觉醒状态，它为脑的其他几个功能系统的各种活动提供基础，而脑的其他几个功能系统的活动则会影响脑的第一功能系统的功能。

心理活动的认知成分和其他心理成分之间的相互作用，主要是通过脑的第二功能系统和脑的其他几个功能系统之间的相互作用来实现的。脑的第二功能系统接受、加工和储存信息。信息加工的结果会影响其他几个功能系统的活动，而脑的其他几个功能系统的活动则对脑的第二功能系统的功能有影响。

心理活动的意志成分和其他心理成分之间的相互作用，主要是通过脑的第三功能系统和脑的其他几个功能系统之间的相互作用来实现的。脑的第三功能系统有制定行为程序的功能，还有进行预测和执行行动等功能。它对脑的其他几个功能系统的活动起调节和控制作用，而脑的其他几个功能系统的活动则会影响脑的第三功能系统的功能。

心理活动的情感成分和其他心理成分之间的相互作用，主要是通过脑的第四功能系统和脑的其他几个功能系统之间的相互作用来实现的。脑的第四功能系统有评估信息和产生情绪体验的功能。

评估-情绪功能系统和保证、调节紧张度与觉醒状态的功能系统之间的相互作用表现为：后者为脑的第四功能系统的活动提供基础，而脑的第四功能系统对信息评估的结果，以及由此产生的情绪体验和做出的反应，则会影响调节紧张度和觉醒状态的功能系统的活动。

评估-情绪功能系统和脑的第二功能系统之间的相互作用表现为：接受、加工和储存信息的功能系统为评估-情绪功能系统提供资料，而在接受、加工和储存信息的过程中又不断进行着评估。评估过程涉及对客观事件的感知、对事件意义的解释、对个体过去经验的提取以及事件信息与储存信息之间的比较等。评估-情绪功能系统的评估结果和情绪体验会影响接受、加工和储存信息的过程。

评估-情绪功能系统和脑的第三功能系统之间的相互作用表现为：评估功能系统的评估结果是编制程序、调节和控制的功能活动的前提；评估功能系统对信息的意义进行评估，选择其中对个体有重要意义的信息，送到编制程序、调节和控制的功能系统，指导它完成调控任务，使后者起调节和控制心理活动与行为的作用，达到期望的最终目标；而脑的第三功能系统则影响评估过程的进行，并且进一步改变情绪体验。

在每一种心智活动中都有许多集成过程，下面以感知觉作为例子。人的感知觉是基本的心理过程。外界事物的物理刺激作用于人的感觉器官，转换为神经脉冲，它们由周边神经系统传送到脑，引起相关脑区的激活，当脑区激活水平达到一定值时，产生相应的感觉体验。外界事物的物理刺激是客观的物理事件，身体内部的神经传递和脑区激活是人体的生理活动，人的感觉体验是主观的心理活动。

对于个体来说，感觉器官接受的各种物理刺激都是信息，事物的信息在脑内进行加工。通过脑内的信息加工和意识活动，对有关信息作出解释，形成对信息意义的理解。信息产生主观体验，但信息和主观体验不同。脑内将多种信息综合的过程是信息集成，而通过信息加工和意识活动将多种主观体验形成整体主观体验的过程是心理集成。信息集成是心理集成的基础，但心理集成不同于信息集成，心理集成的主观体验具有心理学的意义。

前面提到，心理学中把感觉定义为对客观事物具体特性的体验，把知觉定义为对客观事物整体特性的体验。在感知觉过程中有心理集成现象，下面举一个视知觉的简单例子：看一个红色小球在一个平台的某处朝一个方向运动。一个物体有各种不同的属性，如物体形状的属性、物体颜色的属性、物体位置的属性、物体运动的属性等。这些不同的属性分别引起不同的主观体验，有对物体形状是小球的主观体验，有对物体颜色是红色的主观体验，有对物体位置是在平台上的主观体验,以及有对物体朝一个方向运动的主观体验等。

人知觉到的并不是物体许多孤立的特性,而是物体的整体特性。也就是说,人把上述各种不同的主观体验集成为一个整体的主观体验。例如,知觉到一个红色小球在平台上朝一个方向运动的整体体验。心理学把人对物体各种不同体验集成为对物体的整体体验的现象称为绑定(binding)。体验绑定的意思是把各种体验捆绑在一起。上面的例子是视知觉的绑定。

人有眼、耳、鼻、舌、身等多种感觉器官,不同的感觉器官接收外界不同的刺激,通过不同的感觉通道将不同刺激产生的神经脉冲传到脑。视觉是人的多种感觉中的一种,此外还有听觉、嗅觉、触觉、味觉等不同感觉。在上面的例子中,如果这个运动的小球还发出各种声响,人会有对发声的运动小球的整体体验。这时除了对视觉通道接收刺激产生的体验集成为视知觉外,同时会将视觉通道刺激产生的体验和听觉通道刺激产生的体验集成起来。在日常生活中,当人同时接收多种刺激时,会把相应的多种感受综合起来。例如,人在看艺术演出时,看到表演的动作,听到表演的音乐,这些视觉和听觉的体验综合起来,进一步形成对演出的感受。这是跨感觉通道的知觉绑定。

绑定是最常见的心理集成现象。心理学中研究得最多的是知觉绑定,实际上在心理活动的许多过程中都存在绑定。心理学家不但讨论知觉的绑定,而且讨论工作记忆的绑定。知觉过程和工作记忆过程不断交互作用,知觉绑定和工作记忆绑定是交织在一起的。

在脑科学中，对绑定的神经机制进行过许多研究。研究者认为，神经系统的信息处理，首先是由不同的神经细胞群分别检测外界刺激的特征，这称为特征检测；然后将各类神经细胞群分别处理的特征集成起来，实现特征绑定（Treisman, Sykes, & Gelade, 1977; Treisman & Gelade, 1980）。例如，客体的视觉信息包括时空信息和表面特征信息两方面，视觉系统有平行的腹侧通路和背侧通路，它们分别将不同的信息传递到顶叶和颞叶脑区进行集成（Underleider & Mishkin, 1982）。

冯·德·马尔斯堡等人（1981）曾提出同步发放假设来解释知觉绑定，认为脑内神经活动存在振荡脉冲。对于一个知觉对象，它的各种不同的特征分别是由不同的神经细胞群检测的，而神经系统振荡脉冲的同步发放则把许多不同的神经细胞群分别加工的信息集成在一起，从而达到对许多种不同特征的绑定，形成统一的知觉。埃克霍恩（Eckhorn）等人（1988）、格雷（Gray）等人（1989）及辛格（Singer）和盖瑞（Gary）（1995）对脑内40Hz振荡进行实验研究，他们的工作为同步发放假设提供了一些初步的证据。

再以语言的集成现象作为例子。语言是人类不同于其他动物的特征之一，它是一种复杂的心智现象（Chomsky, 1957; 吕叔湘, 1979; 索绪尔, 1980）。人理解语言和产生语言，分别包括各种不同性质的集成作用和集成过程。

个体理解语言的过程是：对自己获得的语言材料进行加工，同时提取原有的知识，把它们结合起来，在脑内构建这些语言材料的意义。在整个过程中，加工语言材料、提取原有知识以及构建语言意义等，都需要脑内进行不同性质的集成。因此，理解语言是集成过程。

个体产生语言的过程是：整理自己的思想，确定自己要表达的内容，将思想转换成语言材料，并且输出语言。在整个过程中，整理思想、确定表达内容以及转换成语言等，都需要脑内进行不同性质的集成。因此，产生语言是集成过程。

心理集成的内容丰富多彩，形式多种多样，后面章节还将对心理集成现象的若干方面进行专门的讨论。

第三章　神经集成论、脑集成论和仿脑学

前面两章分别讨论了集成的脑和脑的集成。为了深入研究神经系统和脑的集成现象的特性和规律及其应用，这一章提出，需要建立三门新的学科，即神经集成论、脑集成论和仿脑学。

神经集成论是研究神经系统集成现象的特性和规律及其应用的学科。脑集成论是研究脑内集成现象的特性和规律及其应用的学科。仿脑学是研究仿造脑的学科。

3.1 神经集成论

前面两章着重在脑的系统水平上讨论脑的集成现象。神经系统有许多层次，在神经系统的各个不同层次，都存在不同性质的集成作用和集成过程。谢林顿（1906）曾对神经系统的各种集成作用进行过系统的讨论。

在神经系统的微观水平上，谢林顿讨论过突触的集成作用和单

个神经元的集成作用,认为可以在突触和单个神经元水平来看整体脑的集成作用。在神经系统的较复杂层次,他讨论过脊髓的集成作用,认为中枢神经系统最基本的作用在于它的整合(即集成)作用。

突触的集成作用表现为:在突触处对许多输入进行集成,产生一个输出。谢林顿说,每一个突触是一个协调机构。

单个神经元的集成作用表现为:一个运动神经元对兴奋输入和抑制输入进行集成,从而决定行为。谢林顿说,单个神经元是集成作用的细胞基础。

脊髓的集成作用表现为:脊髓对兴奋过程和抑制过程进行集成,给出整合性的应答。谢林顿说,反射是中枢神经系统的基本活动方式之一,也是中枢神经系统集成作用的单元反应;脊髓反射是对各种输入协调的结果。

谢林顿所做的这些讨论都是关于神经系统不同层次的信号的集成。他说,神经系统集成作用的特点在于,它们不是通过细胞间的物质输运来实现,而是通过神经信号的传导来实现的。神经信号的集成是神经集成的一个重要方面,在这种过程中,集成作用把不同的神经信号集成起来产生总的输出,并使动物体集成为统一的整体。

必须指出,信息集成只是神经系统的一种集成过程,在神经系统中不仅存在大量的信息集成,还有结构集成和功能集成等其他集成过程。例如,神经细胞集成为神经元簇以至神经回路的过程,就既有结构集成,又有功能集成。因此,研究神经系统的集成现象,不但要研究信息集成，还要研究结构集成和功能集成等各种集成过程。

总之，在神经系统的各个层次上，神经系统的各个部分，在它们所处的环境中，通过它们之间的相互作用，以及它们和环境之间的相互作用，组织成为协调活动的统一整体。在神经系统的不同层次，有不同的集成成分、集成作用和集成过程，分别形成不同层次的统一体。神经系统的不同层次都有多种集成过程，包括物质集成、能量集成、结构集成、功能集成、信息集成等。

在考察神经系统不同层次的各种集成作用和集成过程的实验事实的基础上，我们可以构建一门新的学科，把它定名为神经集成论。神经集成论是研究神经系统的集成现象的特性和规律及其应用的学科，它的英文名称是 neuro-integratics。神经集成论的重要概念是神经系统的集成作用和协调活动等。

3.2 脑集成论

在脑的宏观水平，功能专一性脑区、脑功能系统和整体脑等不同层次都存在大量的集成现象。

脑是集成的统一体。脑内不同层次和不同种类的集成成分，基于它们之间的各种相互作用，构成不同层次、不同形式、多种功能的集成体，最后集成为统一的、具有复杂结构和复杂功能的、协调活动的脑，涌现出丰富多彩的心智，并且产生多种多样的行为。

从整体的脑看，脑内有各个功能系统，它们是脑的集成成分，各个功能系统之间的相互作用是脑内的集成作用，脑所处的身体和体外环境是脑的集成环境；脑内的集成过程是脑内功能系统通过脑内集成作用以及脑、身体与环境的集成作用组织成为集成统一体的过程；脑内的集成过程是随时间发展的动态过程。

在脑的不同层次上，存在多种多样的集成过程，如物质集成、能量集成、结构集成、功能集成、信息集成以及心理集成等。在考察这些集成过程的实验事实的基础上，可以构建一门新的学科，来研究脑内集成现象的特性和规律及其应用。我们把这门学科定名为脑集成论，它的英文名称是brain integratics。更确切地说，这个学科可称为脑与心智集成论，它的英文名称是brain-mind integratics。

脑是神经系统的一部分，神经集成论包含脑集成论，但是它们研究的侧重点不同。神经集成论侧重在神经系统的微观水平和介观水平上研究神经系统的集成现象和原理，而脑集成论则侧重在神经系统的宏观水平（即整体脑及心理活动的水平）上研究神经系统的集成现象和原理。

在19世纪以来神经科学发展的历史上，一些神经科学家曾经对脑功能是分区定位还是浑然一体的问题，进行过长期的争论。必须指出的是，我们讨论的脑集成论是在现代科学的基础上提出的理论，它和神经科学历史上出现过的关于脑功能是浑然一体的学说是两回事。

19世纪中期，弗卢朗（Flourens）（1960）提出脑功能的整体论（holistic theory）。20世纪20年代末，拉什利（Lashley）（1929）提出

脑的记忆的整合论（integrationism）。这两种学说是和脑功能定位论（localizationism）相对立的、关于脑功能非定位的两种代表性学说。

这两种学说认为，脑的各种具体功能都是整个大脑皮质浑然一体的功能，它们与整个大脑皮质的所有部分都有关系，而不是大脑皮质某个特殊部位的功能。这两种学说在神经科学历史上有过一定的影响，但是现代脑科学大量的脑功能成像实验已经证明他们当时提出的观点并不全面。

我们讨论的脑集成论和上述两种学说完全不同。脑集成论和上述两种学说对整合一词有完全不同的理解：在脑集成论中，集成（或整合）一词是指脑内有关部分通过集成作用组成统一体，集成是脑内动态的组织过程；而在上述两种学说中，整合一词是"浑然一体"或"功能非定位"的同义词。确切地说，上述两种学说应当称为"脑的等能学说"，而非"脑的整合学说"。

在神经科学发展的历史上，还有过不少与脑内集成现象有关的讨论。除了前面提到的谢林顿（1906）关于中枢神经系统整合作用的研究外，还有鲁利亚（1973）关于脑的功能系统协同活动的观念、特雷斯曼（Treisman）等人（1977；1980）关于注意整合作用的讨论、克里克（Crick）（1984）关于视知觉整合的讨论，以及埃德尔曼（Edelman）和托诺尼（Tononi）（2000）关于脑内整合作用的讨论等。它们都与脑内集成作用有关，但是这些与集成作用有关的研究是比较分散的，并不是关于脑的集成现象的集中的、系统的理论体系。脑集成论的一个任务是把各种与脑内集成现象有关的不同工作集成起来，构建统一的理论体系。

3.3 仿脑学

脑科学研究的问题很多，主要包括探测脑、认识脑、保护脑、开发脑和仿造脑方面（唐孝威等，2006）。

探测脑是要无损伤地测量活体脑的结构和功能，以便深入地认识脑。认识脑是要揭示脑的结构和脑功能的本质，了解脑的工作原理与心智活动的规律。保护脑是要在认识脑的基础上预防和治疗脑的疾病，保护脑的健康。开发脑是要在认识脑的基础上开发脑的潜力，提高人的素质。仿造脑是要在认识脑的基础上开发具有人脑特点的高度智能化的计算机或机器，即"仿脑机器"。

我们提出一门专门研究仿造脑的学科，称它为仿脑学，把它的英文名称定为 brainics，由 "brain" 一词及词尾 "ics" 组成，后者是"具有这种性质"的意思。仿脑学是脑科学及认知科学与数学、信息科学、工程技术科学等学科的交叉学科，它是脑科学的一个分支学科。

自然界中多种多样的生物系统是人类进行科学技术研究时的模仿对象。目前已经有一门仿生学的学科，从事仿生学的学者观察自然界中各种不同的生物系统所具有的许多特殊的性质和本领，研究它们的原理和机制，并且把这些特性和原理应用到一些工程技术领域。他们发明模仿生物系统特性和原理的方法，发展模仿生物系统特性和原理的技术，制造模仿生物系统特性和原理的装置，为科学技术的创新服务。

仿脑学是仿生学的一个部分。仿脑学这门学科的任务是：研究脑的结构和功能以及研究脑的高级功能即心智活动，在人脑和人的心智活动中寻找可以参考的特性和原理，为工程技术提供新的设计思想和工作原理，从而发展模仿人脑的工程技术，如模仿人脑的智能技术和智能机器等。

神经系统，特别是脑以及心智活动中存在各种集成现象，这些集成现象的研究也是仿脑学研究的一个重要方面。神经集成论和脑集成论的理论为仿脑学中关于神经系统的集成研究和脑的集成研究提供了理论基础。在仿脑学研究中，要模仿脑和心智的集成现象的特性和原理，从而发明新的仿脑方法，发展新的仿脑技术和制造新的仿脑装置。

第二篇

一般集成论

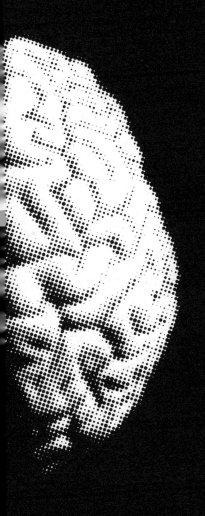

上一篇讨论了脑的集成现象和脑的集成理论。这一篇进一步考察自然界、技术领域和人类社会的集成现象,从实验事实出发,分析不同领域中各种集成作用和集成过程的特性,归纳各种集成现象的一般性概念,构建一般集成论的理论。

这一篇的第四章考察不同领域的一些集成现象;第五章归纳和讨论各种集成现象的一些一般性概念,说明一般集成论的要点;第六章说明一般集成论的特点。

第四章 不同领域的集成现象

在考察了脑的不同层次的集成现象后,我们把目光从脑的领域转向许多其他领域。我们看到,不但在脑的不同层次存在着各种集成现象,而且在自然界、技术领域和人类社会中都广泛地存在着各种集成现象。这一章分别举出自然界、技术领域、学科交叉和人类社会的不同领域中不同性质的集成现象的一些例子。

4.1 自然界的集成现象

在自然界中,无论是无生命的物理世界还是有生命的生物世界,以及高等动物的、以脑活动为物质基础的精神世界,我们都可以看到各种各样的集成过程。

物理世界有层次性结构。物理世界包括微观的物理世界、介观的物理世界、宏观的物理世界和宇观的物理世界。人们在日常生活里接触到的是宏观的物理世界,在宇观的物理世界中有各种天体和空间物质，微观的物理世界中有夸克、电子、原子核、原子等。介观的物理世界指介于宏观的物理世界和微观的物理世界之间的物理世界。例如，一些原子集团（原子团簇），就是介于宏观物质和微观物质之间的介观物质。

在各个层次的物理世界里，有许多集成现象的例子。物理世界中有各种凝聚现象。原子、分子和电子建构成各种集成体，称为凝聚态。凝聚态物质按其空间尺度的不同，有属于介观的物理世界的，也有属于宏观的物理世界的。原子、分子和电子之间的相互作用决定了各种凝聚态物质的内部运动、内部结构、物理性能和外部特征。

在宇观的物理世界中，空间物质形成各种不同的集成体。例如，星系是由恒星和星际物质建构成的集成体。恒星和星际物质之间的相互作用，如引力相互作用和电磁相互作用等，支配了星系内部的运动和结构，决定了星系的各种外部特征。

生物世界的层次性结构表现为：生物世界有生物分子、基因、亚细胞结构、细胞、多细胞生物、生物体器官、生物个体到整个生物界等不同层次。在生物世界的不同层次中，我们都可以看到生物体的各种集成过程。

单个活细胞内部有亚细胞结构和细胞各种成分的集成。活细胞内外有物质运输和能量交换，这些是活细胞和外部环境集成的例子。

多细胞生物体由许多单个细胞集成，这里有细胞的集成，以及多个细胞和环境的集成。从生物有机体个体的内部来看，有结构集成、功能集成和信息集成等各种集成过程。

谢林顿（1906）曾经讨论过动物体内多方面的集成。动物体内各种集成过程的例子有：由大量单个细胞集成为器官，又由各种器官集成为统一的动物个体，这是动物体内结构集成的例子；动物体内各种腺体协同活动，由化学作用实现动物体的集成，还通过血液循环在动物体内传送物质，这是动物体内功能集成的例子；此外，动物体内通过神经系统中神经信号的传导，使分散的器官统一为具有一致性的动物个体，这是动物体内信息集成的例子。

第二章中已经提到，在人的脑和心理活动方面，可以看到许多不同的集成过程。在个体的心理和环境相互作用中，输入是由人的感觉来实现的，输出是由人的运动来实现的。第二章介绍了感觉过程中的集成现象，下面来看运动过程中的集成现象。

人的运动器官能够受心理活动支配而进行各种各样的运动。身体的运动系统和神经系统都参与运动过程。在运动系统方面，有身体的肌肉、骨骼等许多部分的协调活动；在神经系统方面，有脊髓、脑干、丘脑、大脑皮质运动区、小脑、基底节等许多部分的协调活动。

即使是简单的运动，也要由神经系统和运动系统一起实现运动的执行和调控。而复杂的运动，则更和许多因素有关，运动过程中动作种类的选择、运动方向的确定、施力大小的实现以及运动精细程度的控制等，都需要运动的规划部分、运动的执行部分和运动的控制部分的参与，使运动规划、准备、执行、控制等各种功能协调配合。这是人的运动集成的过程。

以上一些例子说明，在物理世界、生物世界和精神世界中都存在集成现象。

4.2 技术领域的集成现象

在工程技术领域中，我们同样可以看到不同种类和不同性质的集成现象。

电子器件在现代技术中有广泛的应用，在这方面，大家都熟知集成元件和集成电路。电子器件和电路的设计和制造极好地展示了集成过程。

光学技术方面的例子是集成光学技术。通信技术方面的例子是互联网集成技术。计算机制造技术方面的例子是计算机集成制造系统，它是计算机技术、电子技术、信息技术、自动控制技术、机械技术和现代管理技术等多种技术集成的智能制造系统。

下面以核物理实验仪器的集成以及神经工程学脑机接口的技术集成作为工程技术领域中集成现象的例子。

科学实验需要实验仪器和实验技术来开展。核物理实验仪器包括各种核探测器和核电子学仪器。在核物理实验中，研究者用核探测器测量和记录核辐射，还用核电子学仪器分析和处理核探测器产生的信号，得到实验数据。核物理实验仪器是由大量的核探测器和核电子学仪器集成的。不同性能的探测元件和探测模块集成为各种核探测器，不同种类的单元电子线路集成为各种核电子学仪器。

电子线路的种类很多，如甄别器、放大器、门电路、扇入与扇出电路、符合线路、反符合线路、脉冲幅度分析器、脉冲时间分析器等。它们分别制备成标准化的单元电子线路。例如，核电子学仪器中的甄别器，是具有对脉冲幅度进行甄别功能的电路。如果输入脉冲信号的幅度超过预设的值（称为甄别阈），那么电子线路就有输出信号；如果输入脉冲信号的幅度达不到甄别阈，那么电子线路就没有输出信号。甄别器是由电子元件按一定规则集成的。又如核电子学仪器中的放大器，是具有将脉冲幅度进行放大功能的电路。一个输入的脉冲信号，经过放大器后，输出的信号幅度增大；这种放大可以是线性放大，也可以是非线性放大。其他各种电子线路也都有各自的功能。这些电子线路都是由电子元件按不同的规则集成的。

各种单元电子线路都有标准化的构件，在使用时根据实际的需要进行逻辑设计，将不同种类的单元电子线路加以集成，再和各种核探测器一起集成为复杂的核物理实验仪器。大规模集成的核物理实验仪器能够对核反应事例进行选择、识别、分析、记录，在核物理实验中起重要的作用。

技术领域中集成现象的另一个例子是神经工程学中脑机接口的技术集成。

神经工程学是神经科学和工程技术科学相结合的交叉学科。脑机接口是用计算机从人心理活动时的脑电中提取信号，利用这些信息和外界环境进行交流。例如，控制机器的运转，并且实现人意图要做的动作（Wolpaw et al., 2002; Wickelgren, 2003）。

人们通常用各种传感器或集成芯片作为与脑接口的器件，由它们在脑和外部设备之间建立交流的通道。在接收和转换人的脑电信号后，计算机驱动的机械装置将代替人的运动器官，并由机器实施动作。人们可以根据需要，分别利用不同的脑电信号来指挥机器进行不同类型的运作，作用于外界环境，以便进行通信或控制等。

在一些实验室中，这种技术被应用在因患病（例如瘫痪）而肢体不能动的患者身上。患者利用有关的装置，就能进行收发电子邮件、开关电器设备、支配轮椅运动等操作，从而在一定程度上与外界环境进行通信，或对机器实施控制。

脑机接口技术涉及许多领域中的不同技术。其中用电极接收脑电信号，就需要有生物学和心理学的实验技术；用计算机处理信号，就需要有信息技术和计算机技术；用计算机驱动机械装置，就需要有计算机技术和机械传动技术等。脑机接口是通过这些技术的集成来实现的。

以上一些例子说明，在工程技术领域中存在着各种集成现象。

4.3 学科交叉的集成现象

现代科学技术发展的一个重要特征是各种不同学科的相互交叉、渗透和融合。在许多具体的科学领域中，深入细致的科学研究使学科高度分化和专业化；同时各个学科的研究需要其他相关学科的参与，通过学科之间的知识集成和技术集成，实现不同学科的交叉研究,并由此产生许多新兴的边缘学科。

学科交叉的例子非常多。以近代生物学和医学的研究为例,它们的发展非常迅速,它们和其他学科之间的交叉研究十分广泛。下面以医学中的医学物理学研究为例子来进行说明。

医学物理学是医学和物理学相结合的交叉学科,它包含的内容很广,其中核医学和放射治疗学是核物理学和核技术在医学应用中的两门学科。

核医学把核技术应用于医学的临床诊断,为人民健康服务。例如,把少量放射性核素或稳定核素标记的诊断药物注入病人体内,在体外用测量放射性核素的射线或稳定核素的特性的影像学仪器,测量诊断药物在病人体内的分布情况和代谢过程,可以达到诊断某些疾病的目的。常用的以18F核素标记的脱氧葡萄糖为诊断药物的正电子发射计算机断层成像技术,在临床诊断恶性肿瘤方面起很大的作用。核医学领域需要核物理学、放射化学、医学、药理学、计算机科学等多种学科的知识集成和技术集成。

放射治疗学把核技术应用于医学的临床治疗,为人民健康服务。例如,用加速器产生射线束,照射病人的肿瘤病灶,杀死肿瘤细胞,达到治疗疾病的目的。现代的质子或重离子治疗用加速器产生质子束或重离子束,其照射范围和照射强度都可以被精密调节,在治疗肿瘤疾病方面起很好的作用。放射治疗学领域需要核物理学、加速器物理学、放射生物学、医学、计算机科学等多种学科的知识集成和技术集成。

在学科交叉中存在许多集成现象。学科交叉是复杂的集成过程，除有不同学科的知识集成和技术集成外，还有合作团队的集成、资源集成和管理集成等。

4.4 人类社会的集成现象

在人类社会中也可以看到许多不同的集成现象。下面以社会团体、社会服务和社会智能为例。

个体是家庭、学校、团体、社会的成员。个体处于家庭、学校、团体、社会等社会环境之中，不能脱离社会环境而孤立存在。社会集成表现为个体集成为集体，在一定的社会环境中生活和工作，并进行社会交往和活动，还受到所处的历史和文化环境的影响。

团队建设是社会中集成现象的一个例子。团队中的成员为着同一个目标，分工合作，互相配合，为社会服务。整个团队要发挥每个成员的积极性，形成和谐团结的集体。这就是团队的集成。

社会的经济、文化、教育等领域有多种多样的集成现象。经济领域有各种经济活动的集成，文化领域有不同文化的集成，教育领域有多元教育的集成。

社会服务集成的一个例子是医疗卫生服务的集成。我国正在建立农村三级医疗卫生服务网，并建设为农村服务的人口与健康科学数据共享平台。农村医疗卫生服务网的建立，是医疗卫生服务机构的集成。人口与健康科学数据共享平台的建设，使农村人口共享有关的科学数据资源，从而提高农村人口整体的健康水平服务，这是医疗卫生信息资源的集成。

再以社会智能集成为例。人类智能活动具有社会性。人类社会中的许多智能活动是在由许多个体集成的集体活动中实现的，因此就有社会智能集成。

戴汝为（2009）曾对社会智能进行过深入的研究。他指出："当今的社会是人类群体和以计算机为代表的机器群体共栖的社会，是人类与其赖以生存的自然界共存的社会，是人类各种思想意识碰撞以及各种思维方式互相激励、演化并进的社会。"他还说："创新，其主体是社会的人、人的群体以及科学团队，表现在工程创新上，它涉及众多学科和工程领域，当然首先体现为对工程创新团队的依赖。"

上面几节中举了自然界、技术领域、学科交叉和人类社会中集成现象的几个例子。虽然例子数目不多，但是可以看到不同领域中集成现象的普遍性。

在自然界、技术领域、学科交叉和人类社会中，诸如原子的凝聚、动物的反射、电路的设计、学科的交叉、团队的合作等一些从表面上看来不相关的现象，具有某些共同的特性。从集成的观点看来，它们是不同种类和不同性质的集成现象。例如，原子凝聚是物理集成，动物反射是神经集成，电路设计是器件集成，学科交叉是知识集成，团队合作是社会集成，等等。因为它们的集成成分、集成作用、集成环境、集成过程等各不相同，所以形成的集成统一体也是千差万别的，且表现出多种多样的集成特性。它们分别属于不同的领域，各自有自身的规律，但是它们都体现了集成现象的共同特点。

用集成的观点考察不同领域中各种不同的集成现象，有助于概括它们的一般概念和一般特性，了解它们的一般原理和一般规律，也有助于应用这些概念和规律去研究与集成现象有关的复杂事物，处理与集成现象有关的复杂事件。

第五章 探索一般集成论

前面几章分别介绍了集成的脑、脑的集成，以及许多不同领域中的集成现象。我们可以看到，在自然界、技术领域和人类社会中，集成现象是广泛存在的。

这一章根据大量的实验事实提出一般集成论的理论。我们先讨论各种集成现象的共同特性，说明一般集成论的要点，再归纳各种集成现象的一些一般性的概念。

5.1 一般集成论理论

集成是过程，是大量集成成分基于它们之间的相互作用建构具有新功能的集成统一体的过程。我们对集成现象的研究是在向脑学习的基础上发展的；这些集成过程不仅在脑内存在，而且在自然界、技术领域和人类社会中广泛存在。

如前所述，在自然界中有大量的、不同层次的集成作用和集成过程。在自然界，包括物理世界、生物世界和精神世界，多种多样的事物组成不同层次的、多种多样的集成统一体；它们分别具有不同的性质。从人的精神世界来看，人的心智活动中存在许多不同的集成过程，人的意识也是在脑功能集成过程中产生的。人类社会活动有各种集成过程，人类个体间相互作用，组成集体和社会。

不同的集成成分及其相互作用具有各自的特性，不同种类的集成过程也各有特殊的性质和不同的规律，需要对它们分别进行具体的分析和研究。而从一般的集成过程来说，各种集成过程有着共同的特性，并且涉及一些相同的概念。一般集成论要考察各个不同领域中的各种集成作用和集成过程，并且通过综合研究，找出它们的共同特性和规律；再从这些一般特性和规律出发，讨论它们在各个具体领域中的应用。

人们早就有集成的观念，对集成或整合并不生疏，在许多不同场合都提到集成或整合，如集成电路、集装箱等。但是在一般集成论中，集成或整合则更有意义。

各种集成现象的共同特点是什么？这个问题需要通过专门的研究来回答。我们的任务是把存在于自然界、技术领域和人类社会中的各种集成现象汇集在一起，把各种集成现象当作专门的科学研究的对象，建立一门新的学科，对它们进行专门的研究。

一般集成论指出，集成现象是复杂系统的普遍现象。在集成过程中，许多集成成分在一定环境中通过它们之间的相互作用以及它们和环境之间的相互作用，组织成为协调活动的统一整体。

集成是一个动态过程。集成统一体是一个整体。集成统一体内的许多成分称为集成成分，集成统一体内的相互作用称为集成作用，集成过程发生的环境称为集成环境，集成成分组织成为集成统一体的过程称为集成过程，集成过程的产物称为集成统一体。

集成过程常有大量集成成分参与。不同种类的集成成分及其相互作用是集成过程的基础。集成成分是参与集成过程并组成集成统一体的单元。复杂系统内部不是单一成分，它们是由多种成分集成的统一体。一些复杂系统具有层次性结构。在每一层次，都有不同的集成作用、集成过程和不同的集成统一体。

集成成分有许多不同的种类。前面提到，在物理世界和生物世界中，集成成分有物质、能量、结构、功能、信息等，因而集成过程有物质集成、能量集成、结构集成、功能集成、信息集成等。精神世界和人类社会中还有其他各种集成过程。

在日常生活中，人们对集成的理解较多侧重在结构集成方面，如集成电路和集装箱等。一般集成论不仅研究结构集成，而且研究物质集成、能量集成、信息集成等。

如前所述，向脑学习为一般集成过程的研究提供了丰富的资料。因为脑和心智的研究不但是脑的结构与功能的研究，还是生理、心理和病理的研究，其中包括主观体验、认知、情感、意志、意识和行为的研究。所以脑的集成过程既有物质集成、能量集成、结构集成、功能集成、信息集成，又有心理集成、行为集成，以至脑和心智与社会的集成。

集成不是集成成分的简单堆积。集成过程的进行要以集成成分之间的相互作用为基础，彼此毫无相互作用的成分是不会进行集成的。集成过程是在一定环境中进行的,系统内部的集成成分通过内部的集成作用以及和环境的相互作用集成为统一体。

集成是一种发展过程,大量的集成成分是在这个动态的发展过程中构建成为具有新功能的集成统一体的。我们可以用一些参量来描述集成过程的特性,如集成度(集成的程度)和集成速度(集成过程的速度)等。

以神经系统为例,托诺尼等人(1994)曾经讨论过神经系统的整合程度$I(X)$。系统X由n个单元x_i组成,各个独立组成单元的熵是$H(x_i)$,系统X作为整体的熵是$H(X)$。

他们把系统X的整合程度定义为所有$H(x_i)$之总和与$H(X)$之差,即:

$$I(X) = \sum_{i=1}^{n} H(x^2) - H(X)$$

$I(X)$表示由组成单元的相互作用导致的熵的减少。若组成单元间的相互作用越强,则$I(X)$的值越大。

在集成过程中,集成体的集成度提高,并在一定条件下展现新现象,使集成统一体出现原来成分并不具有的新的特性,这被称为涌现(emergence)。

总之,集成过程是通过多种多样的集成成分之间各种不同的相互作用实现的。集成成分和相互作用具有多样性,因此会存在不同类型和多种形式的集成过程,它们具有各自的特点;集成过程中形成不同层次和不同特性的模块和网络,最后产生集成统一体。不同类型和多种形式的集成过程,形成千差万别的集成统一体。在集成统一体内部，各个部分在集成作用下协调地活动。

鉴于自然界、技术领域和人类社会中广泛存在各种集成现象的事实，我们认为有必要建立一门称为一般集成论的学科，来专门研究集成现象。一般集成论是一门研究自然界、技术领域和人类社会中各种集成现象的一般特性和规律及其应用的学科。这门学科不仅研究集成作用和集成过程的一般特性和规律，而且探讨如何依据事物本身的性质有效地进行集成和创新的方法。

冯·贝塔朗菲（von Bertalanffy）（1950；1976）把他研究的系统论称为一般系统论（general system theory），因为他讨论的不是某类特定的系统，而是普遍存在于自然界和人类社会中的一般系统。同样地，我们在一般集成论中讨论的不是某类特定的集成现象，而是普遍存在于自然界、技术领域和人类社会中的一般性集成现象。因此我们把所研究的理论称为一般集成论。

我们把一般集成论的英文名称命名为general integratics。选择这个名词是借鉴了信息学的英文名称。信息学是研究信息（information）的科学，英文名称是informatics。一般集成论研究集成（integration）现象，因此命名为integratics。

一般集成论作为一门学科，具有确定的研究对象、研究目标、研究内容和核心概念。

关于研究对象，一般集成论以自然界、技术领域和人类社会中不同层次和不同性质的集成现象为研究对象，从大量集成现象的事实出发，概括它们的共同特征。

关于研究目标，一般集成论以建立一门新的学科为目标，这门学科研究各种集成现象的一般特性和规律；还要将一般集成论应用于自然界、科学技术和人类社会的有关领域，分别研究各个具体领域中集成现象的特性和规律，从而建立一个研究各类集成现象的学科群。

关于研究内容，一般集成论以各种集成现象的共性作为主要的研究内容，着重研究不同领域中不同层次和不同种类的集成现象的共同特性和共同概念，并且在同一个学科中把集成现象的共同特性和共同概念汇集起来，进行综合的研究。

关于核心概念，一般集成论的主要概念是集成。对各种集成现象，都要考察其集成成分、集成作用、集成过程和集成统一体。要讨论物质集成、能量集成、结构集成、功能集成、信息集成、心理集成、知识集成、环境集成、社会集成等，还可以归纳许多集成现象的共同概念，如全局、全局化、模块、模块化、还原、合理还原、综合、有机整合、绑定、联合、联想、建构、重建、优化、临界、涌现、互补、协调、符合、同步、和谐、流畅、适应、同化、顺应、集大成、大统一等。在本章后面几节中，我们将分别讨论这些与集成现象有关的概念。

这里要说明，一般集成论和数学中的集合论是两回事。集合论（set theory）是数学的一个分支（齐纳，约翰逊，1986；方嘉琳，1982）。在集合论中，把凡是具有某种性质的、确定的、有区别的事物的全体称为一个集合（set）。这个数学分支不考虑构成集合的事物的特殊性质，只研究集合本身的性质。

集合论中集合的概念和一般集成论中集成的概念不同。数学中的集合是数学概念,强调数的汇集;而一般集成论中的集成指自然界、技术领城和人类社会中的各种集成现象,特别是其中的集成作用和集成过程。但集成的概念和集合的概念既有区别又有联系。因为集成统一体是包括集成成分的全体,所以集成概念和集合概念也有联系。

集合论的理论和一般集成论的理论是不同的理论。集合论是研究集合的数学性质的数学分支;而一般集成论则是讨论自然界、技术领域和人类社会中集成现象及其规律的学科,着重研究这些集成现象的特性,特别是集成作用和集成过程的特性。当然,在一般集成论的研究中,研究者可以利用集合论中相关的一些数学工具。

一般集成论是关于集成现象一般规律的理论,它为我们提供了观察世界和研究事物的一种观点,也为我们提供了处理事件和解决问题的一种方法。

集成不仅是一般性原理,而且是观察世界和研究事物的观点。既然集成现象是在自然界、技术领城和人类社会中普遍存在的,我们就要用集成的观点去观察和研究那些包含集成现象的各种复杂事物。

对于复杂的事物,我们要从多个方面考察它们所包含的各种集成现象,特别是其中的集成成分、集成作用、集成过程和形成的集成统一体。例如,研究一种复杂的生物体,不仅要考察生物体内的物质集成、能量集成、结构集成、功能集成和信息集成,而且要考察生物体与其他物体的集成,以及生物体与环境的集成。

结构集成、功能集成、信息集成等各类集成过程都是复杂的过程。对于特定的集成过程，要考察哪些集成成分参与这种集成过程，这些集成成分有哪些相互作用，这些相互作用有哪些特性，这种集成过程内部的具体机制是什么，等等。这里涉及集成现象中绑定、建构等许多概念。

集成过程是动力学过程。对于特定的集成过程，我们要考察与这种过程有关的一系列时间特性和集成动力学问题，如集成过程的时间特征是怎样的，集成过程中集成统一体的组织结构是怎样随着时间变化的，集成过程中新的功能是在哪些条件下以及怎样出现的，等等。这里涉及集成现象中同步、涌现等许多概念。

对于复杂的集成统一体，我们都要将其看作是它内部的各种成分通过集成作用而形成的集成统一体。我们要考察集成统一体内部各种成分是怎样相互作用的，集成统一体的各部分是怎样互相配合和协同运行的，等等。这里涉及集成现象中互补、协调等许多概念。

集成不仅是观察世界和研究事物的观点，而且是处理事件和解决问题的方法。既然集成过程在自然界、技术领域和人类社会中广泛存在，我们就要应用集成的方法去处理和解决那些包含集成现象的各种复杂问题。

集成是将分散的各种成分构建为集成统一体的方法。实现集成的一个问题是：如何对具有各种特性的成分进行有效的集成而构建成高效的集成统一体。在许多集成过程中，我们往往根据全局的目标，构筑不同层次和不同性质的模块和网络，最后产生集成统一的产物或输出。这里涉及集成现象中模块化、全局化、优化等概念。

研究复杂事物时，我们常常面临如何分析和还原，以及如何联系和综合等问题。一般集成论提供的方法是合理还原和有机整合的方法，即对复杂事物各部分进行合理的还原，分别对它们进行深入的研究，再根据它们固有的联系与作用，对它们进行有机的整合。这里涉及集成现象中还原、合理还原、综合、有机整合等概念。

处理复杂事件和解决复杂问题，我们要先分析、讨论，再评估、决策，最后组织、实施。在这些过程中，我们都可以运用一般集成论的方法。例如：在分析过程中要掌握全面情况，对各种信息进行集成，得到正确的认识；在决策过程中要集思广益，对各种意见进行集成，形成妥善的方案；在组织工作中要合理配置，对各个部门进行集成，组建协调的团队；在实施过程中要统一指挥，对各个步骤进行集成，达到圆满的结果。

对于包含集成过程的各种事物和事件，应用一般集成论的观点和方法有助于我们理解和解决相关集成过程的许多实际问题。当然，在不同的具体领域中，各种集成过程是各不相同的，所以我们要对不同的具体的集成过程分别进行具体的研究。后面第三篇将讨论一般集成论在各种具体领域中的一些应用。

5.2 全局和模块

与集成现象有关的一组概念是全局和全局化。全局指整个局面，全局化是统筹整个局面的意思。

在集成过程中，我们首先要从全局要求出发，确定集成目标；还要统筹全局，进行全面的集成设计；然后总揽全局，实现集成过程。全局化不只是结构集成的概念，它对于功能集成和信息集成也是重要概念。

心理学家在讨论意识模型时，曾经有过"全局工作空间"的观念（Baars, 1983; Baars & Franklin, 2003）。"全局工作空间"模型认为脑内存在许多专门处理器和一个全局工作空间；专门处理器是专一性的处理器，处理各种信息，可以独立地工作；而全局工作空间则接收各个专门处理器的信息，信息一旦在全局工作空间中得到表达，就可以为理性行为所访问，从而形成意识。全局工作空间可以把意识的内容"广播"到广泛分布于大脑的神经系统，意识的作用是把分散而独立的各种脑功能整合起来。我们曾根据实验事实，对"全局工作空间"模型加以扩展，提出"意识全局工作空间的扩展理论"（宋晓兰，唐孝威，2008）。

与集成现象有关的另一组概念是模块和模块化。模块指标准组件，模块化是在结构集成中制成标准组件再进行组装的意思。

在建构复杂集成体时，我们可以根据全局目标，将集成过程分为几个步骤来完成。先建构各种模块（它们是中间层次的集成体），然后将多个模块集成为复杂集成体。

认知神经科学中有过模块说的学说（Matthei & Fodor, 1984），认为脑是由高度专门化并且相对独立的模块所组成，这些模块经过复杂而巧妙的结合，是实现复杂而精细的认知功能的基础。

心理学中讨论短时记忆时有组块（chunk）的概念。短时记忆的信息容量有限,通常是7±2项,但是可以利用已有的知识经验将信息形成组块。通过扩大组块内的信息容量,人们能够增加短时记忆的总的信息容量。

模块概念不仅适用于结构集成,而且与功能集成和信息集成有关,也适用于工程技术集成以及团体集成与社会集成。

5.3 还原和综合

在集成过程中,许多集成成分通过集成作用而形成集成统一体。还原和综合是与集成现象有关的概念。

还原的意思是分析统一体中的集成成分。综合的意思是将集成成分集成为统一体。还原和综合在集成过程中都是不可缺少的。

前面章节5.1已经提到合理还原和有机整合的方法,《意识论——意识问题的自然科学研究》(唐孝威,2004)一书中讨论过用合理还原和有机整合的方法研究复杂的事物,下面是该书用这种方法考察意识问题的一些说明:

> 首先,要用还原方法研究意识。意识是由多种要素所组成的,还原方法就是要把它分解为其组成的要素以及这些要素之间的相互作用,并对它们分别进行具体的研究。从这个意义上说,意识是应该而且是可以进行还原的。
>
> 其次,这种还原应当是适度的,还原的结果要有心理学

意义。分析意识的基本要素以及要素之间的相互作用，因为它们是基本的，又是具有心理学意义的，所以这种还原是合理的还原。对意识的研究不必还原到单个神经元离子通道的层次。

用合理还原方法研究意识，只是意识研究的一个方面。意识研究的另一个方面，是用有机整合方法研究意识。同时要把合理还原方法和有机整合方法结合起来。

有机整合方法是在对复杂事物进行合理还原的基础上，进一步了解还原要素之间的有机联系，再把各个要素有机地结合起来，得到对复杂事物的整体认识。这种整合是在合理还原基础上的有机整合。

5.4 绑定和联合

与集成现象有关的一个概念是绑定（binding）。前面第二章已经提到，绑定是心理学研究知觉问题时提出的概念，绑定的意思是捆在一起。

以图形识别为例，一个图形有许多不同方面的特征，包括图中的点和线、角度、朝向以及图形的运动等。在图形识别时，视觉系统对图形的许多特征进行检测，检测器测的是各种分散的特征。然而实际上，视觉系统能够把一个图形的各种特征结合到一起，得到对图形的整体知觉。脑是怎样把图形的不同的特征捆在一起的呢？

这被称为特征绑定问题。特征绑定也是知觉集成。

目前认为,注意机制在特征绑定中起关键作用。如果没有注意参与,那么特征是分离的;如果有注意参与,那么脑能够对特征进行绑定,从而知觉到整体的事物(Treisman, Sykes, & Gelade, 1977; Treisman & Gelade, 1980)。

绑定的原来意思是知觉中的特征捆绑,可以把这个概念加以扩展。例如,指在信息集成或知识集成时,把不同信息或知识捆绑在一起。

与集成现象有关的另一组概念是联合和联想。联合指联系与合并,联想是概念联合的意思。

在联合的各种成分中常常有拮抗的方面。例如,神经系统的输入有兴奋作用的和抑制作用的。神经系统的集成作用是把各种输入综合起来,产生总的应答。

在心理学中,联想和粘合是脑内认知活动的方式。脑内关于事物的形象称为表象,联想是指从一件事想到另一件事,这时对原有表象进行综合而产生新的表象;粘合是指把同一件事的各种特征结合而产生新的表象。

脑不仅能按一定方式对检测到的事物的各种特征进行集成而形成一定的认知结构,而且还能根据脑内储存的记忆对事物进行解释和预测。

联想是心理活动中概念集成的一种方法,它可以由脑内的一种概念引起其他概念,或者把不同的概念联系起来,由简单观念结合成复杂观念;或者在不同的观念之间找出它们的关系。联想在心理集成中具有重要作用，也是知识集成的一种有效方式。

5.5 重建和优化

集成现象的一组重要概念是建构和重建。集成过程是系统不断地建构和重建的过程。建构指构造与建设，重建是重新建构的意思。

集成过程具有曲折性。在集成过程中，系统常常根据需要去粗取精，删除多余的及不合适的部分，并对初步结构进行重建。例如，在神经系统发育过程中，系统会对无用部分加以删除，被称为神经元连接的修剪（pruning）。

再以生物进化为例：生物是进化而来的，现今的生物体是在生物长期进化中经过自然选择作用的结果。雅各布（Jacob）（1977）曾把进化比喻成一个"修补匠"。

我们在建构理论时也需要用到重建的观点和方法。克里克在《狂热的追求》（克里克，1994）一书中谈过对建构生物学理论的体会。他说："如果以为只要有一个机智的念头同想象中的事实能稍稍联系起来就可以产生有用的理论，那是相当靠不住的。认为第一次尝试就能做出好的理论，更是不可能的。在获得最大成功之前，他们必须一个接着一个地提出理论。正是放弃一种理论而采用另一种理论的过程，使得他们具有批判性、不偏不倚的态度，这对于他们的成功几乎是必不可少的。"

下面说明集成过程的另一个重要概念：优化。优化是使得尽可能完善的意思。

集成过程有集成目标的优化、集成部件的优化、集成方案的优化、集成方法的优化等。集成目标的优化是集成过程以完成最佳的集成体为目标，集成部件的优化是选择尽可能完善的部件来进行结构集成，集成方案的优化是设计最佳的方案来进行结构与功能集成，集成方法的优化是选择尽可能完善的方法和途径来实现集成过程，等等。对于自然界自发的集成过程，存在自然选择。对于人进行的集成过程，存在人的主动选择，通常要通过评估、试验、比较、选择来达到这些优化。

优化的概念不仅适用于结构与功能集成，它对于各种集成过程都是重要的。人们在集成过程中，都要求优化的和有效的集成，优化的集成也是有效的集成。

5.6 临界和涌现

与集成过程有关的另一些概念是临界和涌现。

在集成过程中，系统不断发生量变。当集成过程中的量变达到一定程度时，系统出现质变，系统在临界条件下会出现新的特性。涌现是集成过程在一定条件下，系统发生质变，出现原来各成分并不具备的新特性的现象。临界条件是系统集成过程中由量变发生质变,并涌现新特性所需要的条件。

《意识论——意识问题的自然科学研究》(唐孝威,2004)一书曾对临界条件进行过如下的说明:

在日常生活中,我们常常见到物理学的物态和物态之间的物理相变。物理学中用相来描述物质的状态。物理学中的物质有不同的状态。例如,同一种物质的气态、液态和固态,就是物质不同的状态或物相。

不同物相是物质的不同状态,气态、液态和固态的物相,分别相当于气相、液相和固相。物质从一种物相到另一种物相的变化称为相变。当物质没有达到相变的临界条件时,物质状态保持一种相,而在达到相变的临界条件时发生相变,物质状态从一种相转变为另一种相。例如,从气相转变为液相,从液相转变为固相等。

在非平衡系统中物理相变的一个例子是:在膨胀云室的气体中,在带电粒子的径迹上凝结水珠,这是从气相转变为液相的相变。另一个例子是:在泡室的过热液体中,在带电粒子的径迹上形成气泡,这是从液相转变为气相的相变。

这本书还对意识涌现现象进行过如下的讨论:

意识涌现是脑功能活动的重要现象。当大脑皮质的给定脑区激活而激活水平还没有达到意识涌现的临界条件即意识阈值时，这个脑区的信息加工是无意识的；而当脑区的激活水平达到意识涌现的临界条件即意识阈值时，就会发生突变，这个脑区的信息加工由无意识加工转变为有意识加工。这时意识涌现，同时产生相应于这个脑区激活态的主观体验。

意识的涌现是突变，从无意识加工转变到有意识加工，或从有意识加工转变到无意识加工，都是不连续的。一定脑区的活动，要么进入意识，要么不进入意识，不是模棱两可的。脑是开放的非平衡系统，脑内意识涌现的现象类似物理学中对称性破缺的情况。

在意识涌现过程中发生许多现象，其中包括：大脑皮质给定的脑区的激活和其他脑区的激活，给定脑区和其他激活脑区之间竞争注意资源，在注意作用下给定脑区的激活受到增强而其他脑区的激活受到抑制，给定脑区激活水平增高超过意识阈值而使脑区信息加工进入意识等。这些现象之间存在紧密的联系。意识涌现过程是竞争资源、选择与淘汰的动态过程。

5.7 互补和协调

与集成过程有关的一个概念是互补性。互补是相互补充的意思。

我国古代思想家早就有互补的观念，如人们熟悉的阴阳互根的观念和天人合一的观念。

在物理学中，波尔（Bohr）提出互补原理（又译为并协原理）。这个原理在讨论物理世界微观现象的波粒二象性时指出，光和实物都有波粒二象性，为了描述微观的物理现象，波和粒子两种概念都是不可缺少的。这两种概念是协调的，从这个意义上说，它们是互补的。

互补原理认为，在描述微观的物理现象时，一些经典概念的描述与另一些经典概念的描述虽然是相互排斥的，但是对全面地描述微观现象而言是不可缺少的；只有把所有这些既排斥又互相补充的概念联合起来，才能够全面地描述微观现象。举一个常见的例子，对一枚硬币来说，正反两面都是不可缺少的，只有正反两面都被看到，人们才能对这枚硬币有全面的认识（玻尔，1964）。

上面所说的互补主要指不同概念的互相补充。在一般集成论中，我们除保留这种对互补的理解外，还把它的意义扩展为功能集成或知识集成时各部分优势的互补，如资源优势的互补及理论观点的互补等。

我们在提出心理学的统一研究取向时指出，当代心理学中各种研究取向之所以不同，并不是因为它们的观点根本对立，而是因为它们关心的问题、考察的重点，以及研究的观点和方法有差别。我们认为各种不同研究取向的某些概念是可以互相补充的，因而可以把不同的研究取向统一起来，从而提出心理学的统一研究取向（唐孝威，2007）。

在集成过程中，我们要把具有互补性的各个方面有效地集成起来，达到优势互补的完美局面。

与集成现象有关的一个概念是协调。协调是配合适当、和谐一致的意思。

前面提到，脑的整体功能是通过脑内几个功能系统来实现的，这些脑功能系统既有分工又有整合，人的心理和行为是脑的几个功能系统协同活动的结果。这是功能协调得很好的例子。这些脑功能系统配合适当、和谐一致的活动，保证了人正常的心理和行为。反之，若脑功能系统活动失调，则会导致人的心理和行为的反常，就需要加以调整（Buzsáki，2006）。

协调不但是功能集成的概念，同样是与工程技术集成、管理集成以至社会集成等有关的概念。例如，工程集成中大量机器的有效运转、社会集成中和谐团体的建构等，都涉及协调的概念。

5.8 符合和同步

这里说明一个与集成过程有关的概念：符合。符合的意思是彼此一致。

符合记录是核物理实验中选择记录同时性信号的电子学方法，相应的电子线路称为符合线路。例如，二重符合线路是具有符合功能的二通道的电子线路。只有当两个输入信号在一个短的时间间隔（称为符合分辨时间）之内输入时，才有符合输出；如果两个信号在符合分辨时间以外输入，那么就没有符合输出。同样还有多通道的多重符合线路。

与此相关的反符合线路是具有反符合功能的电子线路。只有当从符合通道有输入信号，同时不存在从反符合通道来的输入信号时，才有脉冲输出；如果那时从反符合通道有输入信号，那么就没有脉冲输出。又如，延迟符合线路是具有延迟符合功能的电子线路。只有当两个输入信号中的一个信号在某一时刻输入，而另一个信号在这时刻之后指定的一段时间（称为延迟时间）的分辨时间之内输入时，才有延迟符合输出。

在信息集成中，符合是指集成过程中在时间上选择信号的方法，用这种方法可以选择记录时间上相同的信号，也可以记录不同空间位置的同时发生的事件。可以把这个意思加以扩展、总结，解释为集成过程中对有关许多方面选择其共同点的方法。这在信息集成或知识集成时都是适用的。

再看与集成过程有关的另一个概念：同步。同步的意思是一起变动。

同步加速是高能加速器加速粒子的一种原理和方法。同步加速器是加速高能粒子的大型实验装置。同步辐射装置是利用电子同步加速器产生光辐射的大型实验装置。

在高能物理实验中对粒子进行同步加速的原理是，用高频电场不断加速带电粒子，随着粒子速度增加和粒子质量由相对论效应增加，将加速器所用的磁场强度也随着时间相应地增加，这样就可以保持带电粒子在半径为恒定的环形轨道上被加速。

同步辐射光是电子同步加速器中电子在磁场中做曲线运动时产

生的光辐射，通常可以产生包括红光、可见光和X光波段的光。这种大型实验装置被称为同步辐射光源装置。

同步可以指在功能集成或信息集成时各种相关功能在时间上一起变动。也可以把这个概念扩展为在功能集成或信息集成时各个部分同步地协调互动，系统内部各部分有序地同步活动，保证了系统的和谐稳定。

集成过程的另一个概念是流畅性。流畅是物质运输、能量传递或信息流动畅通的意思。

前面介绍过脑内信息加工的组织层次，有信息加工组织的上面层次和信息加工组织的下面层次。在信息加工组织的上面层次和信息加工组织的下面层次之间不断进行信息流动，由信息加工组织上面层次传向信息加工组织下面层次的信息流动是自上而下的信息流，由信息加工组织下面层次传向信息加工组织上面层次的信息流动是自下而上的信息流。信息加工过程是自下而上的信息流和自上而下的信息流交互作用的过程。

信息加工时信息通过信息通道流动，信息通道的畅通是有效地进行信息加工的重要条件。当一个激活脑区对它连接的脑区作用时，它们之间连接通路的每次导通都会使连接通道的效能有所提高，效能提高的程度随着连接通路导通次数增多而增加。这被称为通道的易化（facilitation）（唐孝威，2003a）。

流畅性原来是指在信息集成时信息流的畅通，如今我们也可以把这个概念扩展为结构集成或功能集成的动态过程中的有效连接。

5.9 适应和同化

适应是与环境集成有关的一个概念。适应的意思是适合环境。

各种事物都不是孤立存在，而是处于它周围的环境之中。环境是变化的，事物要不断调整，以适应变化环境的条件和需要。事物和环境的有效集成，要求事物和环境友好共处，形成有效地互动的统一体。

认知科学中曾经有过情境认知的观念（Brooks, 1991），认为认知置身于情境中，依赖于环境；认知具有现场情境的特点，认知过程必须置身现场情境；认知过程还依赖现场情境，认知活动和现场情境密切联系，不能分开。

适应不仅是环境集成的概念，还是生物集成和社会集成的概念。

与环境集成有关的另外两个概念是同化（assimilation）和顺应（accommodation）。同化和顺应是皮亚杰（Piaget）（1983）研究认知结构时提出的概念。在心理学中，同化和顺应都是认知适应的方式。同化是把新信息纳入已有的认知结构之中，顺应是改变已有的认知结构来适应新的环境和信息。

皮亚杰的发生认识论理论指出，人在认识世界的过程中与环境相互作用，形成自己的认知结构，称为图式。一种图式结构经过同化、顺应、平衡而构成新的图式结构。这样，个体通过与环境相互作用而不断适应变化着的环境。

同化和顺应的概念不但适用于环境集成过程的讨论，而且可以把这两个概念扩展为在理论集成或社会集成时的相互适应。

5.10 集大成和大统一

与大规模集成过程有关的概念是集大成和大统一。

中国古代思想中就有集大成和大统一的概念。从集成过程的观点看来，集大成是大规模的集成过程，是将所有各种成分集成为大的统一体的过程。大统一是集大成的结果，是对所有各种成分集成而构成全面的、大的统一体。

在构建思想体系时，需要有集大成和大统一的观点和方法。《统一框架下的心理学与认知理论》（唐孝威，2007）一书提出心理相互作用的大统一理论和心理学统一研究取向时，曾经用集大成和大统一的概念进行过讨论。

在阐述心理相互作用的大统一理论时，这本书中是这样说的：

> 这个理论称为心理相互作用的大统一的理论，因为这个理论既分辨其中各种心理相互作用的不同特性，又指出它们具有共同的、统一的基础。这个理论是涵盖所有各种心理相互作用的统一理论，所以我们称它是心理相互作用的大统一理论。

在阐述心理学的统一研究取向时，这本书中是这样说的：

某一种心理现象往往涉及多种心理相互作用。这些心理相互作用并不是彼此无关,而是互相联系的,因此要用多种心理相互作用联系和统一的观点来考察这种心理现象。

心理学的统一研究取向全面地研究各种不同的心理相互作用,并着重于这些心理相互作用的大统一。因此它的研究领域包含心理学的各个领域,比目前心理学的各种不同的研究取向关心的领域更加广泛。它可以把这些不同的研究取向的主要的、合理的观点统一起来,在取其精华的基础上,集当代心理学各种不同研究取向之大成。

第六章 一般集成论的特点

第一篇考察了脑的实验事实,并且提出了脑集成论的理论。这一篇的前两章从脑的集成现象的研究出发,扩展到一般集成论的研究。

这一章阐述一般集成论的特点,强调一般集成论是在向脑学习的基础上发展的理论,提出可以将一般集成论的理论应用于各种具体领域,构建各种专门集成论的理论,还说明一般集成论与系统论等理论的联系和区别。

6.1 在向脑学习基础上发展的一般集成论

20世纪80年代,有学者提出,研究思维科学有两种途径:一种途径是脑科学的途径,研究脑,弄清人类思维时脑的活动机制;另一种途径是人工智能的途径,寻找人的思维规律,用计算机来模拟实现人脑的功能,把思维科学的研究同人工智能、智能机的工作结合起来。当时研究者认为，脑科学的研究途径太遥远，因此要选择人工智能的研究途径（钱学森，1986）。

现在的情况已经不同了。脑科学实验技术的迅速发展，为人类思维的实质性研究提供了各种条件，从脑机制的角度入手进行思维的脑科学研究是切实可行的（唐孝威，2003a）。我们认为，思维是脑的高级功能，要了解思维，不能离开脑。因此思维科学的研究应当把上面提到的第一种研究途径和第二种研究途径结合起来，即把脑科学的研究和人工智能的研究结合起来，而且必须强调以脑研究为基础，进行思维的脑机制的实验和理论研究（唐孝威等，2006）。

不但思维科学的研究是这样，一般集成论的研究也是这样。研究一般集成论要从脑的研究入手，考察脑的集成现象，学习脑的集成原理。因为脑是自然界最复杂的物质，脑的活动是自然界最复杂的运动形式，脑的结构和脑的活动为我们提供了非常丰富的与集成现象有关的实验资料。

一般集成论理论的特点之一是向脑学习。一般集成论的研究是从脑的集成现象的研究开始的，在向脑学习的基础上，发展一般集成论的理论。脑内存在多种多样的集成作用和集成过程，研究和学习脑内的集成作用和集成过程，有助于发展一般集成论的观念和构建一般集成论的理论。

在研究自然现象，包括无生命的物理世界和有生命的生物世界时，如果不考察脑和心智的活动，那么通常讨论的主要是物理世界和生物世界中物质运动、能量转换、结构连接、功能调节、信息交流等过程，而不涉及人的主观的心智活动。这些讨论涉及的概念主要是物质、能量、结构、功能和信息等。

向脑学习使我们对世界的认识更加广阔和丰富了。在研究脑的结构和脑的活动时，我们不但面对物质的脑和生物的脑，包括脑的物理现象和生命现象，而且还面对心智和行为，即精神世界以及心智、脑和身体与所处的自然环境和社会环境之间的相互作用。

脑和心智现象涉及物理世界、生物世界和精神世界。考察脑和心智活动时，我们不但要讨论脑内多种多样的物质运动、能量转换、结构连接、功能调节和信息交流等过程，而且要讨论丰富多彩的主观体验、认知活动、情感活动、意志活动等以人为主体的精神活动和行为活动，以及与精神活动和行为活动相关的人类社会活动。因此，对脑和心智现象的研究涉及的概念，不仅是与其他生命活动共有的概念，如物质、能量、结构、功能、信息等概念，还有一般生命活动没有涉及的许多其他概念，如体验、认知、情感、意志、意识、行为等概念。

我们提出的一般集成论是从脑科学的研究出发，在向脑学习的基础上，再发展一般集成论的理论。在研究脑的集成现象时，除了考察脑内的物质集成、能量集成、结构集成、功能集成、信息集成等现象外，与脑和心智活动有关的主观体验、认知、情感、意志，以至人类行为等现象，也为我们提供了大量的精神领域、行为领域和社会领域中集成过程的例子。

神经集成论、脑集成论和心理集成论是脑与心智领域中的理论，它们是属于脑科学和心理学等具体领域中的专门理论。我们的目的不但是建立神经集成论、脑集成论和心理集成论，而且要从脑集成论等出发，进一步发展一般集成论的理论。

为了构建一般集成论的理论，我们就要在向脑学习的基础上扩展视野，考察自然界、技术领域和人类社会中的各种集成作用和集成过程。一般集成论是研究不同领域中普遍存在的各种集成作用和集成过程的一般特性和共同规律的理论。

6.2 一般集成论与专门集成论

如前所述，一般集成论的讨论是从脑科学的研究成果出发，再进一步扩展到考察不同领域的集成现象。第四章中举出的一些例子说明，在不同领域中集成现象是广泛存在的。一般集成论面对的是广泛存在于自然界、技术领域和人类社会中的各种不同的集成现象。

上一节讨论了一般集成论的一个特点，即向脑学习；这一节要说明一般集成论的另一个特点，即研究各类不同的集成现象的共性。

一般集成论作为一门学科，它不仅要讨论特定领域中大量的集成现象，还要说明不同领域中各种集成作用和集成过程的广泛存在，并且要研究不同种类的集成作用和集成过程的共同特性。它讨论的概念不仅是一些特定集成过程的相关概念，而且要说明不同领域中各种集成过程的共同特性和一般概念。

我们在第五章中考察了不同种类的集成现象的一般特性，归纳了各种集成现象的一般性概念，阐述了有关集成现象研究的一般性观点，把我们提出的集成理论称为一般集成论。在研究一般集成论的基础上，我们还要用一般集成论的观点研究各种不同的领域，考察各个不同领域中的集成作用和集成过程的特性和规律及其应用。将一般集成论应用于各种专门领域，就要发展研究不同领域中集成现象的一系列子学科。我们把这些子学科称为各种专门集成论，英文名称是 special integratics。

一般集成论和专门集成论既有区别又有联系。一般集成论讨论不同领域中普遍存在的各种集成作用和集成过程的一般特性，一般集成论也是广义的集成理论；而专门集成论则讨论各个专门领域中集成作用和集成过程的具体特性，专门集成论是具体领域的集成理论。原则上说，一般集成论包含各种专门集成论的共同特性和共同规律。

过去有学者曾讨论过个别的具体领域中集成（即整合）的理论，并提出过令人感兴趣的观点。本书的内容和这些工作不同，本书主要研究一般集成论，即在自然界、技术领域和人类社会中广泛存在的各种集成现象的一般特性，因此不同于某些个别的具体领域中集成的理论。本书着重考察脑内的集成作用和集成过程，是为了在此基础上发展一般集成论的理论。

某个具体领域的专门集成论，是将一般集成论应用于这个具体领域，研究这个具体领域中的集成现象的特性和它们的具体规律及其应用而构建的学科。

在自然界的各个具体领域的专门集成论的例子有：研究生物领

域中集成现象的特性和规律及其应用的生物集成论，研究细胞的集成现象的特性和规律及其应用的细胞集成论，研究神经系统集成现象的特性和规律及其应用的神经集成论，研究脑内集成现象的特性和规律及其应用的脑集成论，研究人体集成现象的特性和规律及其应用的人体集成论，研究医学领域中集成现象的特性和规律及其应用的医学集成论，研究心理领域中集成现象的特性和规律及其应用的心理集成论，研究认知领域中集成现象的特性和规律及其应用的认知集成论，研究智能领域中集成现象的特性和规律及其应用的智能集成论，研究环境科学领域中集成现象的特性和规律及其应用的环境集成论，研究地球科学领域中集成现象的特性和规律及其应用的地球集成论，研究空间科学领域中集成现象的特性和规律及其应用的空间集成论，等等。

在工程技术的一些具体领域的专门集成论的例子有：研究信息科学领域中集成现象的特性和规律及其应用的信息集成论，研究工程领域中集成现象的特性和规律及其应用的工程集成论，研究技术领域中集成现象的特性和规律及其应用的技术集成论，等等。

在人类社会的一些具体领域的专门集成论的例子有：研究文化领域中集成现象的特性和规律及其应用的文化集成论，研究艺术领域中集成现象的特性和规律及其应用的艺术集成论，研究教育领域中集成现象的特性和规律及其应用的教育集成论，研究经济领域中集成现象的特性和规律及其应用的经济集成论，研究管理领域中集成现象的特性和规律及其应用的管理集成论，研究社会领域中集成现象的特性和规律及其应用的社会集成论，等等。

这些专门集成论构成了一个庞大的学科群，一般集成论是这个学科群中各个学科的集成。在后面第三篇中，我们将在专门集成论的学科群中选择若干具体领域，分别对这几个具体领域的专门集成论进行讨论。

近年来，集成现象作为一种复杂性现象受到一些学者的关注。他们曾分别对一些领域中的集成现象进行过讨论，提出了各自的观点。下面介绍其中一部分工作（因为收集到的资料不全，列举的工作可能有遗漏）。

刘晓强（1997）认为，集成论的研究对象包括两个方面：一是各种集成，如信息集成、技术集成、系统集成、功能集成、过程集成、环境集成、人与组织的集成等；二是各种集成之间的相互作用。他认为，集成论的研究内容包括以下几个方面：集成的分类、形式、产生条件和形成机制，集成的原理、规律和方法，各种集成之间的关系。他还提出了人的集成、综合集成、复杂巨系统分析、建模与仿真、集成论基础研究等若干研究方向。

海峰、李必强、冯艳飞（2001）在对广泛存在于经济和社会组织中的集成现象进行分析的基础上，以系统论为基础，分析集成的本质和内涵，研究集成论的基本问题和基本范畴。他们认为，集成论的研究目标在于探讨包括集成条件、集成机理和集成规律等在内的集成理论体系，提出了集成单元、集成模式、集成界面、集成条件、集成环境等集成理论的基本问题和基本范畴。

此外，有学者研究过一些具体领域中的集成现象，提出了相应的理论。例如，牛世盛（1997）关于生命整合论的研究，海峰、李必强（1999）关于管理集成论的研究，胡启勇（2002）关于文化整合论的研究，包含飞（2003）关于生物医学知识整合论的研究，李必强和胡浩（2004）关于企业产权集成论的研究，喻红阳、李海婴、吕鑫（2005）关于网络组织集成论的研究，陈捷娜、吴秋明（2007）关于产业集群的集成论的研究，等等。

上述一些研究工作分别从不同角度讨论了各种集成现象。虽然这些工作中并没有提出一般集成论的体系，但是他们进行了调查、分析和讨论，提供了集成现象的多方面的资料，为进一步的一般集成论研究做出了贡献。

6.3 一般集成论与系统论等理论的联系和区别

在 20 世纪，冯·贝塔朗菲的一般系统论、维纳（Wiener）的控制论、香农（Shannon）和韦弗（Weaver）的信息论、皮亚杰的结构主义理论，钱学森的开放的复杂巨系统理论陆续发表。这些理论在系统、控制、信息、结构、开放的复杂巨系统等方面进行了许多研究，阐述了系统、控制、信息、结构主义、综合集成法等许多重要概念。这些理论对 20 世纪科学技术产生了重大的、广泛的影响，也对本书一般集成论的研究有重要的启示。

这一节说明一般集成论和上述几种理论的联系，以及一般集成论区别于这些理论的特点。

冯·贝塔朗菲等人对一般系统论进行了长期的研究。他的代表作是《一般系统论——基础·发展·应用》（贝塔朗菲，1987）。

一般系统论把系统定义为有相互关系的元素的集合。它着重研究适用于一般系统的模型和原理。一般系统论指出，一个系统的主要特征是整体性，整体大于部分之和。

一般集成论的理论和一般系统论有联系，在一般集成论的理论中有集成的系统和集成统一体等概念。但一般集成论讨论的问题和一般系统论讨论的问题是不同的，一般集成论着重讨论集成现象，特别是集成作用和集成过程的特性，而非讨论一个系统的特性。一般集成论强调集成是过程，要用过程的观点研究集成现象。

维纳等人对控制论进行了系统性的研究。他的代表作是《控制论——或关于在动物和机器中控制和通信的科学》（维纳，2007）。

控制论着重研究系统和环境之间以及系统内部的信息交换和通信过程，特别是其中自动调节的特性。控制论提出的主要概念是控制和反馈，认为系统对于环境的功能控制是通过反馈来实现的，反馈机制是动物和机器的有目的的行为的基础。

一般集成论的理论和控制论有联系。一般集成论的理论有集成过程的调节和控制等概念。但是集成过程非常复杂，调节和控制等特性只是集成过程的一部分特性。一般集成论讨论的问题和控制论讨论的问题是不同的，一般集成论讨论集成现象，不但涉及集成过程的调节和控制，而且研究集成过程的其他许多机制，如互补、同步、优化、涌现等。

香农和韦弗对信息论进行了深入的研究。他们的代表作是《通信的数学理论》（Shannon & Weaver，1949）。

信息论的主要概念是信息，用与热力学负熵相应的表示来定义信息。信息论讨论了信息的接受、传送、编码和利用等问题。

一般集成论的理论和信息论有联系。一般集成论的理论讨论集成过程中的信息加工，而且研究集成现象中的信息集成。但是信息加工和信息集成只是一般集成论面对的多种多样的集成现象中的一类过程。一般集成论不仅讨论信息集成过程，而且讨论结构集成、功能集成、心理集成、社会集成等许多其他种类的集成过程。

皮亚杰讨论科学认识的结构主义。他的代表作《结构主义》（皮亚杰，2006）一书是他的"发生认识论"的一个组成部分。

结构主义的观念可以追溯到由德·索绪尔（de Saussure）提出的语言学中关于语言的共时性的系统的概念（索绪尔，1980），以及心理学中完形学派的感知场概念（Hothersall，1984）。

结构主义理论考察了许多不同学科领域的科学认识的结构，如数学结构、物理学结构、生物学结构、心理学结构、语言学结构、结构在社会研究中的应用，以及结构主义和哲学等。结构主义认为结构有三个特性：整体性、转换性和自身调整性。

一般集成论的理论和结构主义理论有联系，一般集成论的理论讨论结构问题，特别是集成现象中的结构集成和理论集成。但是一般集成论考察的范围很广，不仅是结构集成和理论集成，更不限于科学认识的结构。

钱学森在开放的复杂巨系统的理论方面进行了开创性的工作。他的代表性著作有《关于思维科学》（钱学森，1986）、《创建系统学》（钱学森，2007）等。

他研究人工智能系统，提出开放的复杂巨系统的"从定性到定量的综合集成方法（metasynthesis）"。这个方法把专家群体、数据和信息与计算机硬件、软件技术有机地结合起来，使这三者本身构成一个系统。

在此基础上他还进一步提出"从定性到定量的综合集成研讨厅体系（hall for workshop of metasynthetic engineering）"。这是采取人机结合、以人为主的技术路线，把世界上千百万人的聪明才能，包括人的思维、思维的成果、学科的知识、经验知识、各种情报资料及各种有关的信息综合起来，从而解决复杂问题。这些思想对指导科学技术，特别是人与机器结合的智能系统的未来发展具有深远的意义。

一般集成论的理论和开放的复杂巨系统理论有联系，两者都涉及集成的概念。但一般集成论的理论和开放的复杂巨系统理论所讨论的问题是不同的。开放的复杂巨系统理论从工程技术的角度，着重研究人工智能特别是人机结合、以人为主的智能系统，以及从定性到定量的综合集成法；而一般集成论则在向脑学习的基础上发展起来，着重讨论集成现象，特别是集成作用和集成过程的一般特性和规律及其应用。

总之，一般集成论和上述几种理论有许多联系，但一般集成论是和上述几种理论不同的理论。一般集成论和上述这些理论的区别，并不是它们的观点互相排斥，而是一般集成论和这些理论研究的对象、研究的内容、讨论的概念、研究的方法和应用的范围等都有所不同。

从研究的对象来说，上述这些理论的研究对象分别是一般系统、控制过程、信息原理、知识结构、开放的复杂巨系统等，而一般集成论则把自然界、技术领域和人类社会中广泛存在的集成现象作为研究对象。

从研究的内容来说，上述这些理论分别研究系统、控制、信息、结构、综合集成法等方面的规律，而一般集成论则研究自然界、技术领域和人类社会中各种集成现象的一般特性和规律，还分别研究各种类型的结构集成、功能集成、信息集成、心理集成、知识集成、环境集成的集成作用和集成过程。

从讨论的概念来说，上述这些理论分别讨论与系统、控制、信息、结构、人工智能有关的各种概念，而一般集成论讨论与集成有关的概念。虽然一般集成论的理论也涉及集成过程中的系统、控制、信息等概念，但是一般集成论着重讨论的是集成、协调、优化、互补、同步等概念。

从研究的方法来说，一般集成论和上述这些理论不同。一般集成论研究的出发点是向脑学习，从考察脑内集成现象开始，再进一步发展到对其他领城的集成现象的研究，基于大量实验事实，归纳和总结集成现象的一般特性和规律。但是上述这些理论并不研究脑，并不涉及脑的结构、脑的功能、脑内信息加工、脑的高级活动（例如意识）等。

从应用的范围来说，一般集成论应用于不同的具体领域，讨论各个具体领域中的集成现象，从而发展出一个庞大的子学科群，如生物集成论、医学集成论、心理集成论、工程集成论、技术集成论、教育集成论、社会集成论等。这和上述这些理论也是不同的。

总之,一般集成论和上述这些理论既有联系又有区别。一般集成论的理论和前人理论是可以互相补充的,希望一般集成论的研究以及一系列专门集成论的研究能够在前人理论的基础上起补充和丰富前人理论的作用。

第三篇 一般集成论的应用

将一般集成论的理论应用到不同的领域，就要分别研究不同领域中的集成现象及其特点与规律。例如，研究生物集成、心理集成、技术集成、工程集成、教育集成、社会集成等，这些研究形成一系列相关的学科。在下面几章中，我们将以生物集成论、心理集成论、知识集成论、工程集成论和教育集成论为例对一般集成论的应用进行讨论。

另一个很大的研究领域是将一般集成论理论应用于社会领域，研究社会领域的各种集成现象。我们把研究社会领域集成现象和规律及其应用的学科称为社会集成论，它的英文名称是 social integratics。它包括对各种社会现象中集成过程的研究。例如，研究管理工作中集成现象和规律及其应用的管理集成论，研究团队组织中集成现象和规律及其应用的团队集成论等。这方面的研究具有实际意义，第六章已介绍过一些现有的工作，值得在这些工作的基础上，开展系统的、深入的研究（金迪斯，鲍尔斯等，2005）。

第七章　生物集成论

地球上存在多种多样的生物体，它们进行着形形色色的生命活动。第四章已经提到，生物世界具有层次性结构，从生物分子到整个生物界存在许多不同的层次。在不同层次的生命活动中，有各种不同性质的集成现象。

应用一般集成论的观点对生物世界中的各种集成现象进行研究，将有助于了解不同层次生物体及其生命活动的特点。为此，我们提出建立一门研究生物世界集成现象和规律及其应用的学科，并把它命名为生物集成论，它的英文名称是 bio-integratics。

各种生物都是进化的产物。不同层次生物体的集成现象各有特点。因此，生物集成论的研究范围很广。这一章选出三个问题进行讨论，分别是生物集成与生物进化、活细胞的集成和人体的集成。

7.1 生物集成与生物进化

生物世界有许多层次。生物大分子，如核酸和蛋白质，是构成生物体的基本原料，还有许多其他生物大分子，如糖、膜分子等。由生物大分子、离子、水分子等构成的细胞是各种生物体的单位。由许多细胞构成的多细胞生物体，如动物、植物等，具有复杂的结构。生物群体和环境构成生态系统。

不同层次的生物体和生命活动具有不同的空间尺度和时间尺度。例如：生物大分子的空间尺度约是 10^{-8} 米量级，它们活动的时间尺度约在 10^{-9} 秒以下；活细胞的空间尺度约是 10^{-6} 米量级，它们的生命活动的时间尺度约在 10^{-2} 秒以上；多细胞生物体的空间尺度约在 10^{-1} 米以上，它们的生命活动的时间尺度约在 10^{-1} 秒以上；生态系统的空间尺度约在 10^{2} 米以上，它们的生命活动的时间尺度约在 10^{2} 秒以上。

前面第四章提到过生物世界中的集成现象。不同层次的生物体都是集成的产物。例如，生物大分子是小分子集成的，细胞是生物大分子集成的，多细胞生物是细胞集成的，生态系统是生物体和环境集成的（常杰，葛滢，2005）。

对于生物集成现象的研究，特别令人感兴趣的是不同层次生物体不同的集成作用和集成过程。在生物集成过程中，各种生物相互作用起重要的作用。不同的集成元素通过各种生物相互作用实现生物体的集成。在生命活动中存在几种不同性质的生物相互作用，它们是生物体内部成分之间的相互作用，生物体之间的相互作用，生物体和环境之间的相互作用等。这些相互作用具有相互性。例如：在生物体和环境之间，一方面，环境作用于生物体，改变生物体；另一方面，生物体作用于环境，改变环境。

每一个生物体都有内部的生命活动，如代谢、生长等。每一个生物体都处于一定的环境中，并且和其他生物体共存。生物体和环境之间不断有物质、能量、信息的交流。生物体与其他生物体不断进行通信和交流。对于一个生物体来说，既有生物体内部的集成过程，又有生物体外部的集成过程。生物体内部各种成分集成为统一的生物体，这种内部的集成过程是通过生物体内部成分之间的相互作用来实现的。生物体和环境集成为生物与环境的统一体，这种外部的集成过程是通过生物体之间的相互作用及生物体和环境之间的相互作用来实现的。

生物世界中有各种各样的生物体。达尔文的物种起源学说揭示了生物进化和自然选择的规律，指出现今各种生物都经历了长期进化的过程，现代人是从高等灵长类动物进化而来的。从古人猿进化到现代人，经历了大约一千万年的漫长岁月（Eccles，2004）。

现代人的神经系统也是进化的产物。现代人的神经系统具有精致的感觉、知觉和灵巧的运动、控制等功能，还有各种高级功能，它们都是高等灵长类动物经过长期进化和自然选择的结果。

埃克尔斯的《脑的进化》（Eccles，2004）一书在考古学、神经解剖学和脑生理学等方面收集了关于古猿进化到人的大量科学证据，叙述了人类进化过程的主要特征，研究了人类从古猿的脑演变到现代人的脑的进化史，还探讨了高等动物意识的起源。

对于生活在当时地球环境条件下的人类祖先来说，直立和二足行走在进化过程中起重要的作用。例如，在直立和二足行走时，需要大脑皮质对各种肌肉的精细运动进行控制。人类劳动对脑的发展具有决定性的影响。例如，在劳动中制造和使用木制和石制工具，需要手的精巧运动，这促进了大脑皮质运动区及其神经连接的进化。人类言语交流则促使大脑皮质与语言有关区域的进化。总之，脑的结构和功能都在进化中不断发展，人脑是长期进化的结果。

生物进化中有许多集成现象，这些集成现象的特点是经历非常长的时间进程，集成的目标和步骤都不是早先设计好，而是在进化过程中通过生物和环境的相互作用和自然选择逐步实现的。生物进化中的集成过程是"修补"式的集成过程。

7.2 活细胞的集成

细胞是进化的产物。细胞有原核细胞与真核细胞之分，原核细胞是没有细胞核结构的细胞，真核细胞是具有细胞核结构的细胞。这里只讨论真核细胞的情况。

活细胞不断进行着各种生命活动。我们讨论的活细胞的集成，是指活细胞生命活动中的各种集成现象。活细胞有复杂的结构和功能，它们是由各种细胞物质集成的，包括结构集成和功能集成。细胞内部具有四个结构和功能系统，即细胞核系统、细胞质系统、细胞膜系统和细胞骨架系统。细胞内的物质主要有遗传物质、细胞质物质、细胞膜物质、细胞骨架物质等。

细胞核系统是细胞的遗传物质系统，其主要功能是实现遗传。遗传物质是载有遗传信息的DNA分子，它们集中在细胞核内。细胞核具有核膜界面，把细胞核内部物质和细胞质隔开，为遗传物质的储存和复制提供空间。通过核膜，细胞核内外不断进行物质、能量和信息的交流。细胞核是由内部各种成分集成的，在有丝分裂期间，细胞核进行重新集成。

细胞质系统含有各种物质，如生物分子、离子、水分子等，它们集成为许多种类的细胞器以及整个细胞。细胞质系统的主要功能是实现细胞各种生命活动，在这个系统中有物质运输、能量代谢、信息传递等过程。在有丝分裂期间，细胞质重新集成。

细胞膜系统包括质膜、核膜等，细胞内还有许多具膜的结构。这个系统的主要功能是提供分隔物质的空间和内外物质交流的界面。例如，质膜（即细胞膜）形成细胞内部和外部之间的界面，提供细胞内部生命活动的空间，并且促进细胞内外的交流。质膜有脂双层分子，膜上有许多蛋白质，形成受体、离子通道和水通道等，质膜主要是由这些部分集成的。

细胞膜的网架支撑流动膜模型提出，细胞膜包括脂双层和膜骨架两部分；脂双层有双层结构，上面镶嵌着膜蛋白；脂双层底部的膜骨架由骨架蛋白构成，起支撑脂双层流动膜的作用，同时提供膜物质输运的网络（Tang，1998）。

细胞骨架系统的主要成分是微丝、微管、中间纤维等，由它们集成为细胞内的纤丝结构。这个系统的主要功能是提供细胞支撑结构，并且保证细胞内部物质输运。例如，细胞内的颗粒在马达蛋白的作用下，可以沿着细胞骨架进行转运。

实验上曾用水霉细胞作为材料，研究细胞内骨架系统的结构和功能。用电子显微镜和光学显微摄影方法，测量了细胞内纤丝的结构和特性，以及细胞颗粒沿纤丝运动的现象。实验结果表明，在活细胞内部存在着一个由细胞骨架组成的柔性的细胞输运系统（阎隆飞，唐孝威，刘国琴，1994）。

细胞的四个结构和功能系统都是由生物大分子集成的产物，在活细胞内，这四个系统不是孤立存在的，在它们之间有紧密的联系。活细胞的整体是由这四个系统集成的复杂系统，这四个系统的协调运行，保证了活细胞正常的生命活动。

在真核细胞的生长和分裂过程中，存在许多特殊的集成现象。真核细胞有分裂周期，其中包括间期和分裂期。细胞在间期中生长，细胞生长过程中的物质集成是细胞集成多种形式中的一种。在分裂期形成子细胞，子细胞形成过程中的集成现象是细胞集成的另一种形式。这种过程是在母细胞内部复制和物质分配的基础上，子细胞进行建构的集成过程。

真核细胞有丝分裂包括分裂前期、前中期、中期、后期A、后期B等几个阶段。在分裂前期，细胞内组成染色体，并且呈现纺锤体，纺锤体有两个极。在前中期，染色体运动，移向纺锤体的赤道平面。在中期，全部染色体在纺锤体的赤道平面上排列，并往复振荡，然后一起分裂为染色子体。在后期A，染色子体分别向纺锤体的两极运动。在后期B，染色子体到达纺锤体两极附近，同时细胞分裂为子细胞。

有丝分裂中染色体受的力影响它们的运动，而染色体的运动是细胞集成过程的重要部分。我们曾经对染色体受力的特性进行研究，根据有丝分裂后期染色子体向纺锤体两极运动的实验数据，确定使染色子体运动的力并不与染色子体牵引丝的长度成正比，而是保持恒定值（唐孝威，1992）。

细胞集成在有丝分裂过程中进行着。到后期B，当染色子体到达纺锤体两极附近时，子细胞完成整体的集成，子细胞是由染色子体和重新分配的细胞质与质膜等集成的产物。这时还发生子细胞核的重构，这也是子细胞核的集成。细胞分裂的结果是形成两个相对独立的子细胞。

对细胞集成过程的实验研究有很多。例如，实验上曾用爪蟾卵为材料，专门研究过细胞核重建现象（杨宁等，2003）。用扫描原子力显微镜观察到细胞核重建的动态图像。这项研究表明，细胞核重建是细胞集成的过程。

实验上还曾用花粉管为材料，研究细胞顶端生长现象，观察到在花粉管顶端的生长过程中，花粉管内的物质从胞体源源不断地输运到顶端部位，使顶端得以向前生长，这种生长过程不是连续进行的，而是突跳式进行的（唐孝威，刘国琴，1992）。这项研究表明，花粉管顶端生长是细胞集成的过程。

7.3 人体的集成

人体是复杂的生物系统，具有循环系统、呼吸系统、消化系统、神经系统、骨骼系统、生殖系统等许多子系统。循环系统有血液循环的功能，呼吸系统有吸入氧气、排出二氧化碳的功能，消化系统具有消化食物、摄取营养的功能，神经系统有控制调节的功能，骨骼系统有支撑躯体、保证运动等功能，生殖系统有繁殖后代的功能。此外还有内分泌系统和免疫系统。这些子系统协调地活动，组成统一的人体。在人体的生命活动中，人体内部有物质、能量和信息的交流，同时人体还与外部环境进行物质、能量和信息的交换。

从生物集成的观点来看，人体由各个子系统集成，这里不仅有结构的集成，还有功能的集成和信息的集成。人体是一个整体，谢林顿（1906）在《神经系统的整合作用》一书中讨论过将人体各个部分集成为人的整体的机制。他认为人体各部分是通过神经活动、血液循环、内分泌过程等途径集成的，其中神经的集成作用最为重要。

按照中国传统医学，人体内部存在经络。《关于人体经络的一个试探性假说》（唐孝威，沈小雷，何宏建，2008）一文曾经讨论过人体经络。从人体集成的观点来看，人体经络可能是人体内部集成过程的途径之一。下面引用该文的一些文字进行说明：

中国传统医学的长期实践证实针刺穴位的治疗作用。大量临床观察和实验研究资料表明，针灸对机体各个系统、各个器官功能几乎均能发挥多方面、多环节、多水平、多途径的综合调整作用。

针刺穴位的作用，离不开穴位这个针刺的施术部位。传统中医认为，穴位也称腧穴，是人体脏腑经络之气血输注出入于体表的部位。它们不是孤立于体表的点，而是与脏腑组织器官有着密切联系、互相输通的特殊部位，是诊察和治疗疾病的所在。每一个腧穴都与脏腑有特定的联系，这种联系的通道就是经络。

穴位与脏腑组织器官是互相输通的，穴位的输通作用是双向的。在体表的腧穴处针刺或艾灸等能治疗脏腑经络的病症，脏腑的生理状况及病理变化也可通过经络反映在相应的腧穴上。在病理状态下，某些腧穴常会出现特定的变化。例如，胃肠病患者常在足三里、地机等穴出现明显压痛，肺脏病患者常在肺俞、中府等穴出现明显压痛和皮下结节。脏腑病症在相应腧穴的反映，主要是通过经络来完成的，其主要表现有压痛、酸楚、硬结、松陷等。这有助于诊断疾病，并在治疗上帮助选择有效穴位。

针刺穴位治疗相应脏腑病症，脏腑病症也会通过经络在体表相应的穴位上出现异常变化。这种与针刺穴位相关的脏腑，称为"靶点"，穴位与靶点内外相应。这种体表与内脏之间的相关性，就是以经络为基础的。

对于经络的认识及经络学说是在医疗实践中逐步形成并

不断充实和发展的。它有广泛的实践基础，已成为中医理论的重要内容之一，也是针灸理论的依据。经络是具有联系、反应和调整功能的系统，是人体组织结构的重要组成部分，它与脏腑、形体、官窍等组织器官共同构成了人体，又遍布周身，纵横交贯，通过有规律的循行和复杂的网络交会，将人体联系成统一的有机整体。经络系统由经脉和络脉组成，是由经脉和络脉相互联系、彼此衔接而构成的体系。经脉是经络系统中的主干，深而在里，贯通上下，沟通内外。络脉是经脉别出的分支，浅面在表，纵横交错，遍布全身。经络系统密切联系周身的组织和脏器，在生理、病理和防治疾病方面都起着重要的作用。

科学工作者对经络现象及其实质进行了多方面的观察与研究，但对于经络的物质基础至今还没有定论。

人体是由许多系统集成的一个复杂系统，其中神经系统、内分泌系统和免疫系统组成人体内部具有调控功能的神经-内分泌-免疫的整体系统。近年来对神经-内分泌-免疫网络，特别是神经-内分泌-免疫系统调节的分子机制进行过许多研究。这些研究表明，人体内部的神经系统、内分泌系统和免疫系统是一个整体，它们形成神经-内分泌-免疫的整合性网络。

神经-内分泌-免疫系统是心身统一体的重要组成部分。从生理和心理角度说，心身相互作用是通过神经-内分泌-免疫系统的活动来实现的。神经系统包括中枢神经系统和周围神经系统。在内分泌系统中，内分泌腺释放激素，影响体内效应器的活动。激素是内分泌腺分泌的化学物质。肾上腺素、去甲肾上腺素、皮质醇等都是激素。免疫系统产生抗体,抵抗外来的病原。抗体能够识别和抵抗体内异物。例如,血液和体液中的抗体具有杀灭和抑制细菌的功能。

在神经-内分泌-免疫系统中,有神经信号的传递,还有化学物质的传递,包括各种激素和神经递质等。这些化学物质会和细胞受体结合而影响免疫系统的功能。麦卡恩(McCann)等人(1998)和梅尔梅德(Melmed)(2001)的论著,对在神经和内分泌系统与免疫系统界面处发生的生理过程和分子相互作用有详细的阐述。

神经信号可以直接支配体内效应器的活动,又可以通过支配内分泌腺的活动来调节和控制体内效应器的活动。例如,在情绪激动时,脑内信号引起自主神经系统的反应,调控内分泌系统的活动。肾上腺分泌皮质激素,通过血液传送到体内各部分。皮质激素的升高会抑制免疫系统的活动,影响免疫系统抵抗疾病的能力。而免疫系统活动的变化反过来会影响神经系统,迈尔(Maier)和沃特金斯(Watkins)(1998)曾研究过免疫系统对中枢神经系统的作用。

内分泌激素是通过血液流动而传递的。因为化学物质的传递速度比神经信号的传递速度低,所以内分泌的化学物质的作用在时间上比神经信号的作用慢,而内分泌的化学物质作用的持续时间则比神经信号作用的持续时间长。

心理神经免疫学的研究表明:在神经-内分泌-免疫系统

中，如果任何一个系统发生紊乱，那么就会对其他两个系统产生不利的影响。例如，神经系统的紊乱会使内分泌系统失调和免疫系统功能减退。在免疫系统中，若抗体的形成过程受到障碍，则人的免疫功能会失调。

通常认为，神经-内分泌-免疫系统遍及全身，具有弥散分布的特点。虽然目前这种弥散分布的神经-内分泌-免疫系统的观点能够说明许多生理现象，但是它难以直接说明人体具有确定穴位和针刺穴位治疗作用的经验事实。我们对目前弥散分布的神经-内分泌-免疫系统的观点加以改进，提出具有敏感节点和功能连接的神经-内分泌-免疫网络的假设。

自然界中存在的大量复杂系统都可以通过形形色色的网络加以描述。一个典型的网络是由许多节点与连接两个节点之间的一些边组成的，其中节点用来代表真实系统中不同的个体，而边则用来表示个体间的关系。往往是两个节点之间具有某种特定的关系则连一条边，反之则不连边，有边相连的两个节点被看作是相邻的。例如，神经系统可以看做是大量神经细胞通过神经纤维相互连接集成的网络，计算机网络可以看做是自主工作的计算机通过通信介质（如光缆、双绞线、同轴电缆等）相互连接集成的网络。类似的还有电力网络、社会关系网络、食物链网络等。神经-内分泌-免疫系统也是一种复杂的网络。

具有敏感节点和功能连接的神经-内分泌-免疫系统网络假说的要点是：

第一，人体内部的神经-内分泌-免疫系统是一个复杂网络。这个网络既有遍及全身的弥散分布结构，又具有一系列敏感节点和敏感节点间的功能连接。功能连接导致具有空间距离的神经事件之间的时间相关。它区别于结构连接，但又以结构连接为基础，反映了不同结构对神经、生理事件的相似响应。

第二，功能连接是通过神经-内分泌-免疫系统中神经信号的传递和化学物质的传递等物质过程来实现的。

第三，在这个复杂网络的敏感节点上施加的物理刺激（如针刺、艾灸、电刺激等），可以对网络起调控作用。不同的敏感节点与相关靶点相联系，分别起治疗相关疾病的作用。一系列针刺穴位是这个复杂网络的敏感节点。

我们根据人体集成的观点提出的上述假说，把人体经络的实质和具有敏感节点和功能连接的神经-内分泌-免疫系统网络联系起来。如果中国传统医学中的人体经络是具有敏感节点和功能连接的神经-内分泌-免疫系统复杂网络中的组成部分，网络中的敏感节点相当于一系列人体针刺穴位，敏感节点间的功能连接相当于连接一系列针刺穴位的经络，那么人体经络的观念就可以和改进后的神经-内分泌-免疫系统的观念统一起来。当然，这个复杂网络除包含敏感节点和功能连接外，还包含弥散分布的分支通路等。

从生理功能来说，如果人体经络相当于这个复杂网络中一系列敏感节点及其功能连接的部分，人体经络的功能就可以和神经-内分泌-免疫系统的生理功能联系起来。这个复杂网络不仅是遍及全身的系统，而且可以通过敏感节点及其功能连接对身体起调控作用。

复杂网络有空间结构，复杂网络中的相互作用具有时间维度。根据上述假说，对应于针刺复杂网络，针刺穴位的治疗作用既有空间特性，又有时间特性。针刺治疗作用的空间特性表现为：复杂网络中确定的敏感节点有相应的调控靶点并对特定的疾病起治疗作用。针刺治疗作用的时间特性表现为：治疗作用在时间上包含快成分和慢成分两个部分。针刺治疗作用的快成分是针刺穴位引起神经-内分泌-免疫网络激活时神经信号作用的成分。神经信号直接控制靶点，神经信号传递速度快而持续时间较短，因而起即时的调控治疗作用。针刺治疗作用的慢成分是：针刺穴位引起的神经-内分泌-免疫系统网络激活时，内分泌和免疫信号所起的作用。内分泌激素的传递速度比神经信号传递速度慢，但内分泌激素的持续时间比神经信号的持续时间长，因而可能存在治疗的持续效应。

第八章 心理集成论

丰富多彩的心理活动中存在多种多样的集成现象。应用一般集成论的观点，对心理活动中各种集成现象的特点进行研究，是值得关注的。我们提出建立心理集成论学科，这是一门研究心理活动集成现象和规律及其应用的学科，心理集成论的英文名称是psycho-integratics。

心理活动中集成现象非常多。第二章介绍过心理活动中的各种绑定现象，它们是心理集成的典型例子。思维集成也是心理集成的一种表现。这一章将对心理活动集成现象的若干方面进行讨论，比如心理相互作用与心理集成、意识的集成、认知的集成、心智的集成，以及心智与行为的集成等。

8.1 心理相互作用与心理集成

心理相互作用对心理活动的集成过程十分重要，心理集成过程是通过各种心理相互作用来实现的。因此需要研究心理现象中各种心理相互作用的特性，以及如何通过心理相互作用实现不同的心理集成过程。

心理现象包括各种心理和行为。个体的心理现象除内部的心理活动外，还涉及脑、身体、自然环境和社会环境等不同层次的许多因素。心脑、身体、自然环境、社会环境是一个集成的统一体，在这个统一体中，心理活动和脑、身体、环境、社会等各种因素不是孤立的，而是不断地进行着相互作用。

我们受物理学研究各种物理相互作用及其统一性的启发，提出了心理现象中的心理相互作用的概念，在《统一框架下的心理学与认知理论》(唐孝威，2007) 一书中讨论了有关心理现象中心理相互作用的各种问题。我们把心理现象中的相互作用称为心理相互作用，指出在多种多样的心理现象中，存在着几种不同种类的心理相互作用。某种心理相互作用，是心理活动和某种因素之间的相互作用，包括这种因素对心理活动的作用，以及心理活动对这种因素的反作用。以个体的心理活动和社会环境之间的关系为例，社会环境对个体的心理活动有作用，而个体的心理活动对社会环境有反作用。这就是它们之间的相互作用。

丰富多彩的心理活动是在人的脑内进行的，生机勃勃的脑和身体又处在千变万化的环境之中，千变万化的环境中形形色色的事物作用于个体的心、脑和身体，个体的心、脑和身体通过多种多样的行为作用于环境中的事物。因此，在研究个体的心理活动时，既要考察个体的心脑系统内部各种因素之间的关系，又要考察个体的心脑系统与各种外部因素之间的关系。

我们考察心理活动时个体心脑系统内部各种因素的关系，看到有几种不同性质的心理相互作用，其中有心理活动的各种成分之间的相互作用，还有心理活动和脑之间的相互作用。

心理活动包括感觉、知觉、学习、记忆、注意、思维等。心理学中对这些过程都有定义（Sdorow，1995；彭聃龄，2001）。例如，学习是个体在一定情景下由于反复的经验而产生的行为或行为潜能的比较持久的变化，记忆是在头脑中积累和保存个体经验的过程，思维是借助于语言、表象或动作实现的、对客观事物的概括的和间接的认识，推理是由具体事物归纳出一般规律，或由一般原理推出新结论的思维活动。

前面提到，心理活动有觉醒-注意成分、认知成分、情感成分、意志成分等多种成分，这些不同的心理活动成分不是孤立的，而是有紧密联系的，并且不断地进行着相互作用。心理活动的每一种成分和其他成分之间都有相互作用，这些作用具有相互性，即心理活动的这种成分对其他成分有作用，而心理活动的其他成分对这种成分也有作用。以认知活动和情感活动之间的关系为例，认知活动对情感活动有作用，情感活动对认知活动也有作用，这就是它们之间的相互作用。这些相互作用称为心理活动各种成分之间的相互作用，简称心理成分相互作用。心智是各种心理成分的集成，心智的集成是通过心理成分相互作用来实现的。

从心理活动和脑的关系来看，心理活动是脑的功能，脑是心理活动的基础。心理活动的各种成分是脑功能而不是某种实体，所以心理成分相互作用不是一些实体之间的相互作用。然而这些心理成分都以脑为物质基础。心理活动的各种成分之间相互作用的脑机制，是脑内各个功能系统之间的相互作用。

心理活动和脑之间有相互作用，包括心理活动对脑的作用，以及脑对心理活动的作用，这些作用具有相互性。这些相互作用称为心理活动和脑之间的相互作用，简称心脑作用。心脑系统是心智与脑的集成，心脑集成是通过心理成分相互作用和心脑相互作用来实现的。

心脑关系是心理活动和脑之间的关系，心脑相互作用是心脑关系的重要内容。心理活动不能离开脑进行，心脑相互作用是作为脑的功能的心理活动和脑的实体之间的相互作用，因此心脑相互作用的性质和心理成分相互作用的性质不同。

我们再来考察个体心理活动时心脑系统和各种外部因素的关系，可以看到心理活动和多个层次的外部因素之间的各种相互作用，它们是不同性质的心理相互作用。这些作用都具有相互性，既有作用，又有反作用。

从心理活动和身体的关系来看，与心理活动相互联系的各种生理信号在身体内部的神经-内分泌-免疫系统中传递，将心理活动与身体活动联系起来。以神经信号为例，外界环境的刺激在身体的接收器官处产生神经信号，这些神经信号由身体内的神经系统传递到脑，引起各种感知觉；而由脑输出的神经信号则由身体内的神经系统传递到身体的各部分，支配身体运动器官的运动。心理活动和身体之间的相互作用包括心理活动对身体的作用，以及身体对心理活动的作用。这些相互作用被称为心理活动和身体之间的相互作用，简称心身相互作用。心、脑、身体系统是心、脑、身体的集成，这种集成是通过心理成分相互作用、心脑相互作用和心身相互作用来实现的。

心身关系是心理活动和身体之间的关系。脑是身体的器官，是身体的一个部分。心理活动作为脑的功能，不能离开脑而进行，脑又不能离开身体而独立存在。心身相互作用是心身关系的重要内容。心身相互作用的性质和心脑相互作用不同。在心身关系和心身相互作用中，心不是一种实体，而是脑的功能。

个体总是处于外界环境之中，在个体的心理活动和自然环境之间有相互作用。一方面，心理活动通过脑和身体，产生行动而对自然环境作用；另一方面，自然环境不断给个体各种刺激，通过身体和脑而对心理活动作用。这些相互作用称为心理活动和自然环境之间的相互作用，简称心物相互作用。心-脑-身体-环境系统是心、脑、身体、环境的集成，这种集成是通过心理成分相互作用、心脑相互作用、心身相互作用和心物相互作用来实现的。

心物关系是心理活动和环境中物的关系，即个体的心理活动和个体所处的自然环境中的客观事物的关系。心物相互作用是心物关系的重要内容。心物相互作用是由心、脑和身体组成的系统和客观环境之间的相互作用。心物相互作用和心身相互作用的性质不同。在心物关系和心物相互作用中，心不是某种实体，而是脑的功能。

个体处于社会环境中，在个体的心理活动和社会环境之间有相互作用。一方面，心理活动通过脑和身体产生行动而对社会环境作用；另一方面，社会环境不断给个体各种刺激，通过身体和脑而对心理活动作用。这些相互作用称为心理活动和社会环境之间的相互作用，简称心理-社会相互作用。心、脑、身体、社会环境是心、脑、身体、社会环境的集成，这种集成是通过心理成分相互作用、心脑相互作用、心身相互作用、心物相互作用和心理-社会相互作用来实现的。

个体心理活动和社会环境有密切的关系，个体心理-社会相互作用是个体心理和社会环境关系的一个重要内容。心理-社会相互作用和其他心理相互作用的性质不同。在个体心理和社会环境的相互作用中，个体的心理活动不是一种实体，而是个体脑的功能。

上面提到心理活动的各种关系，如心理活动各种成分之间的关系、心理活动和脑之间的关系、心理活动和身体之间的关系、心理活动和环境中物之间的关系以及心理活动和社会环境之间的关系等，在心、脑、身体、自然环境和社会环境集成的统一体中，存在多个层次和多个方面的复杂网络。

归纳起来，心理现象中存在着不同性质的五种心理相互作用，它们是：心理成分相互作用、心脑相互作用、心身相互作用、心物相互作用以及心理-社会相互作用。心理现象中的心理相互作用非常复杂。在一种心理现象中，往往不是只有单独一种心理相互作用，而有多种心理相互作用的集成。在心-脑-身体-自然环境-社会环境的集成统一体中的集成过程，是通过上述五种心理相互作用来实现的。

《智能论——心智能力和行为能力的集成》（唐孝威，2010）一书指出，这几种心理相互作用的种类不同，相互作用的空间范围不同，相互作用的时间范围不同，相互作用的途径不同，相互作用的方式不同，相互作用的结果不同。然而这些不同的心理相互作用都是在心-脑-身体-环境-社会的统一体中的各个层次进行的，它们都以心脑系统的活动作为共同的基础，因而它们可以在心脑统一性的基础上统一起来。

心理相互作用的特点之一是作用的相互性，即既有作用，又有反作用。以心脑相互作用为例，一方面，脑内神经系统的电活动和化学反应是心理活动的生物学基础，它们对心理活动起决定的作用，各种心理活动都有相应的脑机制；另一方面，心理活动过程中伴随着的神经系统的电活动和化学反应，对脑内神经网络起塑造的作用。因此心理活动和脑之间的作用是相互的。

心理相互作用的另一个特点是作用的动态性。个体的脑和心智是在各种心理相互作用的共同作用下发展的。以心脑相互作用为例，脑内不断进行心理活动，脑对心理活动的作用以及心理活动对脑的作用，使心脑系统协调地工作；脑具有可塑性，个体的心理和脑内神经网络在这种动态作用下不断发展。

8.2 意识的集成

意识是复杂的心智现象，意识的集成是复杂的心理集成现象中的一种。对于意识的集成，要从许多不同的方面进行研究。意识状态和意识体验是意识的两个重要方面，这一节从意识状态的集成和意识体验的集成两个方面的集成过程考察意识的集成。

塞尔（Searle）（2000）在讨论意识的特性时说，意识具有一系列性质，如定性的性质、主观的性质、统一的性质和流动的性质等。意识的定性性质是指个体有意识状态时总有一个特定的定性体验，意识的主观性质是指个体有意识状态是个体主观的体验，意识的统一性质是指个体的意识体验是整体的体验，意识的流动性质是指个体的意识体验是随时间不断地更新的。

从意识集成的观点看来，意识的主观性质、统一性质和流动性质都与意识的集成密切相关：意识的整体体验是意识集成的结果，而意识体验的流动性质是意识集成的过程，意识的主观性质则说明意识集成是在个体心脑系统内部的集成现象。

从意识状态的角度考察意识活动，整体的意识状态是由有意识、无意识、潜意识等各种意识状态集成的。心智活动包括有意识活动和无意识活动，心智结构中还有潜意识成分（唐孝威，2008a）。

我们可以按脑内信息加工的内容是否进入个体意识，而把脑内信息加工分为个体有意识的信息加工和个体无意识的信息加工。脑内信息加工过程进入个体意识而被个体觉知的，称为有意识的信息加工，它们参与有意识的认知活动，是外显的信息加工。脑内信息加工过程不进入个体意识而不被个体觉知的，称为无意识的信息加工。例如，脑内有许多信息加工过程，由于加工过程中相应的脑区激活水平低于意识阈值，就不能进入个体意识。虽然无意识的信息加工不被个体觉知，但是它们也参与认知活动，是内隐的信息加工

(唐孝威，2004)。

此外，脑内有些信息加工过程，因为相关的神经活动与觉知系统没有联系，所以它们不可能被个体所觉知。例如，脑内信息的储存过程、脑内信息的传递过程，以及脑内信息加工的步骤等。它们被称为"非意识"的神经活动。脑内专一性的信息是由相应的脑区储存的，当相应的脑区未被激活时，这些信息未被提取，也未被加工。脑内信息处于储存但未被提取加工的状态，称为"潜意识"状态（唐孝威，2008a）。

上面所说的有意识、无意识和潜意识都是意识状态。这里对无意识和潜意识的定义和有些研究者的定义不同，有些研究者把这里说的无意识活动称为潜意识或下意识，而把这里说的潜意识状态称为记忆痕迹。

在《意识论——意识问题的自然科学研究》(唐孝威，2004)一书中，我们把不同的意识状态看作是意识的不同能态：潜意识是意识的基态，无意识活动是脑区激活后水平在意识阈值以下的低激发态，有意识活动是脑区激活水平在意识阈值以上的高激发态。

整体的意识状态是意识的基态、低激发态和高激发态的集成。在一定条件下，这些不同的能态之间会发生跃迁，例如，信息的有意识提取是意识基态到高激发态的跃迁，信息的无意识加工转变为信息的有意识加工是意识低激发态到高激发态的跃迁，信息的有意识加工转变为信息的无意识加工是意识高激发态到低激发态的跃迁，等等。在巴斯(Baars)(1988)提出的意识全局工作空间理论、迪昂(Dehaene)(2001)提出的意识的神经全局工作空间理论和我们提出的扩展的意识全局工作空间理论（宋晓兰，唐孝威，2008）中，都讨论了无意识活动进入全局工作空间而成为有意识活动的过程。

作为复杂心理现象的意识是有内部结构的，意识体验有一些基本的要素，如意识觉醒、意识内容、意识指向和意识情感。大量的实验事实表明，意识与觉醒相联系，意识活动具有内容，意识过程还伴随着意向和情感。这四个意识要素分别反映意识体验的重要特征，它们都有心理学的意义。从意识体验的角度考察意识活动，整体的意识体验是由意识觉醒体验、意识内容体验、意识指向体验和意识情感体验等各种意识体验集成的。我们在《意识论——意识问题的自然科学研究》（唐孝威，2004）中曾对意识要素进行过讨论。

意识觉醒是意识体验的一个要素，意识具有主观性，意识体验是个体的主观体验，而个体在觉醒状态时才会有各种主观体验。意识觉醒有不同程度，它反映一定时刻个体意识体验的强度。个体不但能体验到自己是否觉醒，而且能体验到自己的觉醒程度。因为一定的觉醒状态是意识活动的基础，所以意识觉醒是意识体验的要素之一，这个要素被称为意识觉醒要素。

个体的意识觉醒要素和生理状况有关，受个体生理状况的制约。例如，意识觉醒程度有昼夜的节律性变化，在清醒时和睡眠时，个体的觉醒程度不同，觉醒程度还会随着个体身体生理状况而变化。

意识体验具有整体性，意识觉醒要素描述个体意识体验的整体状况。意识体验具有流动性，个体意识觉醒程度是变动着的，某一时刻个体意识觉醒程度是这一时刻的特性，它反映的是这一时刻附近的一定时间间隔内觉醒程度的平均水平。觉醒是觉知的必要条件，意识觉醒还和意识体验的其他要素有关。

意识内容是意识体验的另一个要素。个体不但能知道自己是否有觉知，而且还能知道自己觉知的是什么或觉知了什么。个体觉知的是什么或觉知了什么，就是个体主观体验的具体内容。个体在这些意识体验的基础上，还能进一步知道自己觉知的内容所具有的意义。

意识内容包括进入意识的事件和知识等。个体意识体验的具体内容是多种多样的。例如，体验某种过程或某种情景；又如，体会某种观念或某种思想。无论个体觉知的是事物、事件还是知识，它们都有具体的内容，这些内容有心理学的意义。因为意识体验总是包含具体的内容，所以意识内容是意识体验的要素之一。

意识是不断变动的，意识内容不断变动，它们表示在某一时间进入个体意识体验的内容，也就是在这一时间个体意识体验到的具体内容。意识内容是信息，意识内容的变动形成脑内的信息流。在意识内容中有脑内加工的各种信息，即心理活动的内容。除脑接收的输入信息和相应的主观体验外，意识体验的内容还包括对信息意义的理解。在意识内容中还有脑发出的输出信息，即支配的动作的内容。

个体的意向和情感都包括信息，它们也构成广义的意识内容，但是意识内容要素讨论的是事物、事件和知识方面的信息，以及对这些信息意义的理解，它们和意向、情感有区别。因此，除意识内容要素之外，把意识指向要素和意识情感要素列为意识体验的另外两个要素。

意识指向是意识体验的一个要素。意识有指向性或意向性。个体的意识体验导致个体有进一步的意向，特别是在了解所体验内容的意义的基础上，个体的意向更加明确。意识内容是不断流动的，意识指向促进意识内容的流动，使这一个时刻的意识内容转到下一个时刻的意识内容。这些意识指向使个体给自己提出各种目标和计划，并且促使个体实现能动的活动。个体这种在指向方面的主观体验具有心理学的意义，所以它们是意识体验的要素之一。

意识情感是意识体验的另一个要素。心理学把时程较短的感情称为情绪，把有长时程、稳定性的感情称为情感。情绪和情感方面的意识体验都具有心理学的意义，这种主观体验构成意识的重要组成部分。它们是普遍存在的，所以是意识体验的要素之一。

意识体验的每一种意识要素都包括许多不同的类别和特征。以意识内容要素为例，前面提到有两类意识内容：一类是主观体验的事物和事件的内容，如事物的特性、事件的情景及时间、空间特性等，还有对它们意义的理解；另一类是主观体验的知识内容，如各种知识概念和规律等，还有对它们意义的理解。

意识体验的每一类内容还包含许多具体的特征。在事物的特性和事件的情景方面，我们可以按感觉通道分为视觉感知的内容、听觉感知的内容、味觉感知的内容、嗅觉感知的内容、触觉感知的内容，等等。意识内容又有单一特征的感知和复合特征的感知，后者是各种不同的感觉通道之间的特征感知集成的结果。人们讨论最多的是视觉感知的内容，被称为视觉意识（Crick & Koch, 2003;

Zeki, 2003),实际上它们只是意识内容的一部分。类似于意识内容要素,意识觉醒、意识指向和意识情感都分别包括各自的类别和特征。例如,意识觉醒要素包括个体不同觉醒程度的特征,意识指向要素包括各种不同指向的特征,意识情感要素包括各种不同情感的特征等。

由于意识具有主观性,意识总是属于个体自我的。与个体自我相联系的意识活动称为自我意识,其中包括四个意识要素中与个体自我相联系的部分。个体的自我意识不但属于个体自我,而且还是涉及个体自我的意识活动。例如,与个体自我相联系的意识觉醒、意识内容、意识指向和意识情感。

在自我意识中,自我包括个体自身当前的存在和活动、个体自我过去的经历和个体自我未来的规划等,又包括个体与环境的关系和个体与他人的关系等。个体通过自己当前的存在和活动、自己过去的经历和自己未来的规划,以及自己和环境的关系与自己和他人的关系等,来确认个体自我。个体具有自我意识,因而能够认识自我以及自我在环境中的位置(唐孝威,2004)。

意识体验的集成观点不仅着重在心理学方面讨论意识要素及其集成,还在脑的系统水平上讨论意识要素的脑基础和意识要素集成的脑机制。意识体验是意识觉醒、意识内容、意识指向和意识情感这四个意识要素的集成,这四个意识要素对于整体意识都是必需的,缺一不可:个体觉醒才有意识体验,意识体验总有内容,意识体验总有指向,意识体验必定带有情感的"色彩"。

整体意识具有统一性。意识四个要素不是机械地并列的,而是通过它们之间的相互作用集成为整体的意识。在四个意识要素之间有密切的联系。意识的四个要素中最基本的要素是意识觉醒要素，它为意识的其他三个要素提供支持，而意识的其他三个要素都会影响意识觉醒。

意识内容要素和意识情感要素之间的密切联系表现为：意识内容经过脑内分析和评估，评估的结果支配意识情感；而意识情感则对脑内信息加工过程起调制作用，从而影响觉知的意识内容。意识内容要素和意识指向要素之间的密切联系表现为：意识内容经过脑内的评估和抉择，形成意识指向；而意识指向则指导信息获取、选择和加工的过程，从而影响觉知的意识内容。意识情感要素和意识指向要素之间的密切联系表现为：意识情感影响意识指向的形成和发展，而意识指向则使意识情感发生转移和变化（唐孝威，2004）。

8.3 认知的集成

认知是心智的重要成分，是人脑的高级功能。认知包括多个层次和各种形式的活动，如前面提到过的感觉、知觉、学习、记忆、注意、思维、语言等。从一般集成论的观点来看，认知是这些不同形式的活动的集成。在认知过程中，这些活动相互依赖，相互影响。它们之间的相互作用往往不仅是某两种活动之间的相互作用，各种不同形式的活动是交叉地进行的，这些活动之间的各种交叉相互作用形成了认知活动的复杂网络。

认知神经科学把认知看作是脑内的信息加工过程，在信息加工中有各种信息的集成。这一节从信息加工和意识活动集成的角度来考察认知过程，也就是说，在认知过程中不但存在着信息加工，而且存在着意识活动，信息加工和意识活动是紧密耦联和集成在一起的。

认知过程常从外界环境获取信息开始，经过脑内的信息加工、主观感受、意义理解、事件评估、形成决策，再主动调控和支配行动，并作用于环境。

个体受客观事物的物理刺激而产生主观感受，这些感受是个体对物理刺激的内容和性质的主观体验。个体对红色物体有红色的主观感受，对声响刺激有声音的主观感受，对自己身体疼痛有疼痛的主观感受，等等；这些主观感受或主观体验是意识活动的基本特性。对物理刺激的主观感受是认知的基础，它们在认知过程中是必不可少的。

个体的认知不但有对客观事物的物理刺激的主观感受，而且还有对物理刺激相关信息的意义的理解。个体根据自己长期积累的经验，对主观感受做出解释，并且把各种相关的信息组织起来。在有红色的主观感受时，个体会对红色的意义做出自己的解释；在有声音的主观感受时，个体会对声音的意义做出自己的解释；在有疼痛的主观感受时，个体会对疼痛的意义做出自己的解释；等等。这些意义理解是意识活动的重要部分。对物理刺激相关意义的理解是认知内容的一部分，它们在认知过程中是必不可少的。

在认知过程中存在评估与抉择。脑内信息加工包括对信息的评估，并且在评估的基础上进行抉择。评估和抉择是意识活动的一部分，在认知过程中，个体通过对信息意义的理解和对相关事件的评估，产生主观意向，这些主观意向使个体进一步选择性地获取新的信息，从而影响认知过程的进展。

个体经过评估和抉择做出的决定，通过脑内调节控制的功能系统，对机体状态进行调控，并对外界环境做出合适的反应，产生行动作用于外界客观事物。个体对认知过程的主动调控是意识活动的另一重要部分，它们在认知过程中也是必不可少的。

总之，在认知过程中脑内不但有信息加工，而且有包括感受、理解、评估、抉择和调控等意识活动，有信息加工和意识活动的集成。信号变换、信息加工、主观感受、意义理解、事件评估、形成决策、主动调控和输出动作，在认知过程中都是必不可少的。简单的认知信息加工观点只关注脑内的信息加工而忽略主观感受、意义理解、事件评估、形成决策、主动调控等意识活动，就不能全面地了解认知。

《统一框架下的心理学与认知理论》（唐孝威，2007）一书以听人说话和视觉图像辨认两种简单的认知事件为例，来说明认知过程中信息加工和意识活动的集成。在听人说话时，脑内对听到的话语信息进行复杂的信息加工，听话人还对说话者的声音、形象及环境等有主观感受，并且提取脑内原来储存的语义知识，从而对听到的话语的意义有所理解，然后做出判断和反应。如果只讨论认知过程的信息加工而不讨论认知过程中的意识活动，就不能全面地描述听人说话这种简单的认知过程。

在进行视觉图像辨认时，脑内除了进行视觉信息的加工外，还存在大量的意识活动。单纯地提取图像的特征信息，并不能深入了解图像的内容，还需要有对信息的意义的理解，以及从记忆中提取已储存的知识进行分析和推测。这些"自上而下"的加工都是意识活动。用简单的信息加工模型说明视觉图像辨认有一定的效果，但是有局限性。如果只讨论认知过程的信息加工而不讨论认知过程中的意识活动，那么就不能全面地描述视觉图像辨认这种简单的认知过程。

总之，脑内信息加工是认知的一个方面，脑内意识活动也是认知的重要方面。在认知过程中，信息加工和意识活动有紧密的耦联，它们集成在一起，而不是独立无关的。我们对传统的认知信息加工模型进行扩展，提出认知的信息加工与意识活动耦联模型，它是对认知的信息加工观点和认知的意识活动观点进行集成的结果。这个模型在传统的信息加工模型的基础上，强调认知过程中意识活动的作用，全面地讨论认知过程中的信息加工与意识活动，以及它们之间的相互作用和耦联（唐孝威，2007）。

认知的信息加工与意识活动耦联模型和传统的认知信息加工模型的区别是：传统的认知信息加工模型不讨论主观的意识活动，在这些认知模型中只有信息和信息加工的概念，而不提意识和意识活动的概念；认知的信息加工与意识活动耦联模型则强调认知过程中意识活动的重要性，以及意识活动和信息加工之间的相互作用，在这个模型中既有信息和信息加工的概念，又有意识和意识活动的概念，还有信息加工与意识活动相互作用与耦联的概念

(唐孝威，2007）。

8.4 心智的集成

上面两节讨论了意识的集成和认知的集成，意识和认知都是心智现象。心理活动包括心智和行为，这一节先讨论心智的集成，下一节再讨论心智和行为的集成。

心智是脑的功能，具有复杂的结构，包括觉醒、认知、情感、意志等成分。心智的觉醒成分和意识密切相关。认知是心智的重要成分之一，认知成分又包括许多不同过程。心智是各种心智成分通过它们之间的相互作用而集成的结果。

心智的集成具有非常广泛的内容。意识的集成和认知的集成是心智集成中的部分内容，心智能力的集成也是心智集成的一部分。心智能力是心理学的重要问题，这一节着重介绍两种智力模型——PASS智力模型和AMPLE智力模型，并从心智能力集成的角度来说明心智的集成。

纳格利里（Naglieri）和戴斯（Das）(1990) 以及戴斯、纳格利里和卡比（Kirby)(1994) 从认知的三个不同层次阐述智力的特征，提出由计划（planning）、注意（attention）、同时性加工（simultaneous）和继时性加工（successive）四个过程组成的智力模型，简称PASS智力模型。这个模型是智力的认知模型，在《认知过程的评估——智力的PASS理论》(Das, Naglieri, & Kirby,

1994)一书中对这个模型有详细的讨论。

　　脑科学的发展对智力研究有很大的影响。PASS智力模型是基于脑科学研究成果的一种智力模型。这个模型的基础是鲁利亚（1966，1973）的脑的三个功能系统学说。第一章已经介绍过鲁利亚阐述的脑的三个功能系统，即保证、调节紧张度和觉醒状态的功能系统，接受、加工和储存信息的功能系统，以及制定程序、调节和控制心理和行为的功能系统。

　　在PASS智力模型中，智力的四个过程是基于三个层次的认知系统，即注意-唤醒系统、信息加工系统和计划系统。注意-唤醒系统是这个模型中整个认知系统的基础，信息加工系统处于整个认知系统的中间层次，同时性加工和继时性加工是信息加工系统的功能，计划系统处于整个认知系统的最高层次。三个系统动态联系，协调合作，保证了智力活动的进行。

　　戴斯根据PASS智力模型设计过相应的智力测验量表。它由四个分量表组成，包括计划的测量、注意的测量、同时性加工的测量与继时性加工的测量四种认知过程的测量，称为戴斯-纳格利里认知评估系统（Das, Naglieri, & Kirby, 1994）。

　　我们认为，心智能力是觉醒、认知、情感、意志等能力集成的结果，PASS智力模型中讨论的能力只是心智能力的一部分，因此根据新的实验事实和对脑功能系统的认识，对PASS智力模型进行扩展，把局限于描述注意-唤醒过程及认知过程的PASS智力模型扩展为全面描述觉、知、情、意各种心智成分的AMPLE智力模型，其中AMPLE是注意（attention）、操纵（manipulation）、计划（planning）、学习（learming）和评估（evaluation）等过程的简称（唐孝威，2008b）。

脑的四个功能系统学说（唐孝威，黄秉宪，2003）是脑的三个功能系统学说（Luria，1973）的扩展。按照这个学说，人的行为和心理活动是通过脑的四个功能系统相互作用和协调活动而实现的。在智力研究中，用脑的四个功能系统的观点来考察智力，自然会基于脑的四个功能系统学说，对基于脑的三个功能系统学说的PASS智力模型进行相应的扩展。我们基于脑的四个功能系统学说，对PASS智力模型进行以下几方面的扩展（唐孝威，2008b）：

第一，把评估-情绪过程列为智力的基本过程之一。PASS智力模型提到过评估，却没有把评估-情绪功能作为脑的重要功能之一来讨论，也没有把评估-情绪过程列为智力的基本过程之一。脑的四个功能系统学说在鲁利亚三个功能系统的基础上，增加了脑的第四个功能系统，即评估-情绪系统。脑的四个功能系统学说在扩展PASS智力模型时，强调了智力活动中评估-情绪过程的重要性。个体脑内存在着先天遗传的评估-情绪结构。脑内对信息进行评估的结果，会引起情绪体验。评估抉择过程是智力的重要组成部分，情绪也对智力有重要作用。

第二，把学习和记忆过程列为智力的重要过程。PASS智力模型在讨论信息编码时也提到短时记忆和长时记忆，但是并没有把学习和记忆列为智力的基本过程。因为学习和记忆对认知有重要作用，所以在扩展模型中强调了认知加工活动中的学习和记忆过程。

第三，对原来PASS智力模型中的信息编码和加工的内容进行修改，用操纵表征的过程作为主要过程，来代替原来PASS智力模型中的同时性加工和继时性加工过程。实际上，操纵表征的过程包含信息的同时性加工和继时性加工的功能。

第四，强调心智是觉醒、认知、情感、意志等各种成分的集成，心智能力是觉醒-注意能力、认知能力、情感能力和意志能力集成的结果。在扩展模型中，注意、操纵、计划、学习、评估等过程集成为总的智力的内容。

对PASS智力模型进行扩展，并没有否定原来的PASS智力模型，而是在保留原来模型的特色的基础上增加新的内容，即保留原来模型中的计划过程和注意过程等内容，增加了评估-情绪过程以及学习-记忆过程等内容。原来的PASS智力模型主要讨论心智活动中的认知过程，它是智力的认知模型；而扩展后的智力模型则不但包括认知，还包括情感和意志，它是一个囊括觉、知、情、意诸成分在内的集成的智力模型。

AMPLE智力模型基于脑的四个功能系统学说，认为智力活动是以下五种过程，即注意过程、操纵过程、计划过程、学习-记忆过程和评估-情绪过程的集成。

第一种过程：注意过程。原来的PASS智力模型已经指出，注意-唤醒是智力的重要过程。智力活动需要个体的唤醒状态，并且需要可控制的注意，来使脑进行有效的工作。注意-唤醒主要是脑的第一功能系统的功能，其相关脑区是脑干网状结构和边缘系统等。对智力活动来说，注意-唤醒系统是智力各种过程的基础，注意-唤醒过程和智力的其他过程之间的相互作用是通过脑的第一功能系统和其他几个功能系统之间的相互作用来实现的。

第二种过程：操纵过程。脑内有信息的心理表征和对心理表征进行的心理操纵。脑内信息加工不但有信息的编码，而且有信息的处理和对信息意义的理解。这些过程都是智力活动的重要内容。对心理表征的操纵，包含信息的同时性加工过程和继时性加工过程。操纵心理表征的过程是脑的第二功能系统的功能之一，其相关脑区是大脑皮质的枕叶、颞叶、顶叶等。

第三种过程：计划过程。原来的PASS智力模型已经强调，计划是重要的智力过程。在智力活动中，需要个体不断进行预测和计划。计划过程是脑的第三功能系统的功能之一，其相关脑区是大脑皮质的额叶等。计划系统和其他系统集成，使智力活动协调进行：计划系统对注意系统起促进或抑制作用，对操纵心理表征的系统进行监控和调节，并且对行为做规划和调整。

第四种过程：学习-记忆过程。学习和记忆是重要的智力过程。学习是个体的认知结构在与环境相互作用中不断建构的过程，智力包括个体在环境中学习的能力。记忆过程是对信息编码、转换、存储和提取的过程。记忆有长时记忆和短时记忆。学习-记忆过程是脑的第二功能系统和第三功能系统的联合功能。学习-记忆过程和其他过程集成而形成完整的智力，如果没有学习和记忆，那么就无所谓智力。

第五种过程：评估-情绪过程。评估和情绪也是重要的智力过程。智力活动需要个体不断对各种信息进行评估和选择。评估的结果导致情绪体验。评估-情绪过程是脑的第四功能系统的功能，其相关脑区是杏仁核、边缘系统和前额叶的一部分。评估-情绪过程和其他过程集成起来，评估过程和情绪过程不但对计划系统和学习-记忆系统有影响，而且对注意-唤醒系统和心理表征的操纵系统有影响。

由以上五种过程集成的智力模型就是包括注意、操纵、计划、学习、评估过程的AMPLE智力模型。这个智力模型认为，人的智力是多元的，上述多种心理过程以及它们之间的相互作用集成为整体的智力。这些心理过程以脑的四个功能系统为基础而在智力活动中集成起来，并且通过脑的四个功能系统之间的相互作用而协调地活动。这个智力模型是基于脑的四个功能系统的模型，脑的这四个功能系统有神经解剖学和神经生理学的基础，因而它不是思辨的模型，而是有实证基础的模型。

PASS智力模型在智力测验方面提出过一系列测量。AMPLE智力模型是PASS智力模型的扩展，因此除保留PASS智力模型原有的测量外，还要增加一系列新的测量内容，包括学习的测量、记忆的测量、评估的测量、情绪的测量等。这些测量和戴斯-纳格利里认知评估系统的测量联合起来，可以构成更加全面的智力测验。

在智力研究领域中，许多学者曾经先后提出过各种智力理论。从20世纪80年代中期以来，有代表性的智力理论除PASS智力模型之外，还有加德纳（Gardner）的多元智力理论，斯滕伯格（Sternberg）的智力三元理论，沙洛维（Salovey）、梅耶（Mayer）和戈尔曼（Goleman）的情绪智力理论以及霍金斯（Hawkins）、布莱克斯利（Blakeslee）的智力理论等。AMPLE智力模型和这几种智力理论有一些共同点，也有许多不同点。

加德纳（1993）的多元智力理论认为智力是多元的，如言语智力、逻辑-数学智力、空间智力、音乐智力、身体运动智力、社交智力、自知智力等。这个理论没有讨论多元智力之间的关联。AMPLE智力模型是基于脑的四个功能系统学说的多元智力模型，它考察的多元智力的内容和加德纳理论的内容有所不同，而且它还强调多种智力过程之间的集成以及它们的脑机制。

斯滕伯格（1985）的智力三元理论包括智力成分亚理论、智力情境亚理论和智力经验亚理论，其中智力成分亚理论认为智力有三种成分，即元成分、操作成分和知识获得成分。它是智力的认知模型，没有讨论情绪等因素，也没有着重考察智力的脑机制。AMPLE智力模型基于脑的四个功能系统学说，除了讨论脑的第二功能系统的认知功能之外，还强调脑的第一功能系统的觉醒功能、脑的第三功能系统的意向功能以及脑的第四功能系统的评估-情绪功能。因此AMPLE智力模型是一个强调多种智力集成的觉、知、情、意诸成分兼备的智力模型。

沙洛维、梅耶（1990）以及戈尔曼（1995）的情绪智力理论专门讨论情绪智力，即与理解、控制和利用情绪相关的智力，但是没有考察认知过程、意向过程等其他心理过程，因而它不是全面的智力理论。而AMPLE智力模型既强调注意、操纵、计划、学习、记忆、评估等过程，又包括情绪因素。

霍金斯、布莱克斯利（2004）提出的智力模型强调，智力的要素是记忆与预测。AMPLE智力模型讨论的智力活动包括了记忆过程和预测与计划等过程，认为它们是智力活动的重要过程，但是指出它们只是智力的部分内容。因为除了记忆与预测外，注意、操纵、计划、学习、评估等过程都是智力过程，所有这些过程集成起来，才构成完整的智力。

8.5 心智与行为的集成

心理活动包括心智和行为。心智是脑的功能，是主观的心理活动；行为是人的反应、动作等过程，是人外部的表现。心智和行为两者有密切的联系，心智和行为关系的一个例子是人的决策和行动：决策是内部的心智活动，行动是外显的行为活动；行动受决策的支配，有了正确的决策，才有正确的行动，而行动的结果又会影响决策。

从一般集成论的观点来看，人的心理是心智与行为的集成。在心智与行为集成的统一体中，并不是两者简单的总和。心智活动和行为活动不断相互作用，并且交织在一起，《智能论——心智能力和行为能力的集成》（唐孝威，2010）一书在讨论智能的本质时指出，智能有内部的心智能力和外部的行为能力，行为能力和心智能力不同，但它们之间密切相关。心智能力是心智活动的特性，它们是心智活动能做哪些事情以及顺利做这些事情的本领；行为能力是行为活动的特性，它们是行为活动能完成哪些任务以及顺利完成这些任务的本领。

《人类的智能》(潘菽，1985) 一书认为，智能包括"智"和"能"两种成分，"智"是人对事物的认识能力，"能"是人的行动能力，其中包括技能和习惯等；"智"和"能"结合在一起，不可分离；这种能力可以以主观的形式存在于脑内，就是"智"，也可以通过人的行动见效于客观，就是"能"。

我们提出广义的智能定义，认为智能是心智能力和行为能力的集成（唐孝威，2010）。心智能力相当于"以主观的形式存在于脑内"的能力，包括觉醒-注意能力、认知能力、情感能力、意志能力等，但并不局限于"对事物的认知能力"；行为能力相当于"通过人的行动见效于客观"的能力，包括操作能力、表达能力、管理能力、社会能力等；心智能力和行为能力集成为整体的智能。

在智力模型方面，上一节介绍了PASS智力模型和AMPLE智力模型，其中AMPLE智力模型提出，智能活动是心智各种能力的集成，因而它是一个囊括了觉、知、情、意等成分的智力模型。虽然AMPLE智力模型已经由智力的认知模型扩展为智力的心智模型，但是它只讨论了智能活动中的心智能力，而没有讨论智能活动中的行为能力，所以它还不是全面的智力模型。实际上在智能过程中，心智活动和行为活动是紧密联系、集成在一起的。后面第十一章将讨论智能集成论，强调智能是心智能力与行为能力的集成，进一步把智力的心智模型扩展为智力的心智与行为模型，形成全面的、集成的智力模型。

这一章讨论了心智活动中集成现象的若干方面，如心理相互作用、意识、认知、心智、心智与行为等方面的集成。从这些讨论可以看到，心理世界的集成现象有许多不同于物理世界中的集成现象和生物世界中的集成现象的特点。

第九章　知识集成论

人类在长期的实践和研究中不断获得关于自然界、工程技术和人类社会等各方面的知识。通过生产实践、工程实践和社会实践,特别是通过近代科学对自然现象的实验研究和理论研究,得到了大量的新知识,使人类知识在广度上和深度上有了前所未有的飞速发展(Wilson, 1998)。

现今人类的知识宝库是人类历史长河中所有知识的集成。数千年来,勤劳智慧的中华民族对人类文明做出过许多贡献,中华文化是人类知识宝库的一部分。

在知识发展中存在知识集成现象,需要应用一般集成论的观点,对知识集成的特点进行专门的研究,总结知识集成的经验和规律。因此我们提出建立知识集成论的学科,这是一门研究知识发展中各种集成现象规律及其应用的学科,其英文名称是knowledge integratics。

这一章先讨论知识集成与学科交叉,然后用几个具体例子说明几种不同形式的知识集成现象,它们是选择性注意的集成模型、心理学学科体系的集成和认知科学理论的集成。

9.1 知识集成与学科交叉

知识的结构具有复杂的层次性,在各个层次上,存在着多种类型的知识集成现象。下面列举知识集成的若干方面,如知识资源的集成、理论模型的集成、学科体系的集成、研究取向的集成、不同学科的集成等。

知识资源的集成是一种知识集成。关于知识资源的集成,大家熟悉的是图书馆和资料室,包括近年来发展的数字图书馆和电子资料室。这里不对它们展开讨论,只着重谈谈基于互联网的科学数据库与科学数据共享。在现代科学技术的一些领域中,总是有许多单位在同一个领域中进行工作,最常见的是临床医学的诊断和治疗。人们在长期实践中积累了大量的科学资料,在各自工作的情况下,这些资料大多是分散的,有些是重复的,但许多是可以互相补充的。需要把这些资料集中起来保存,并加以分类管理。

目前,许多科学技术领域都开展了知识资源集成方面的工作,收集大量与本领域相关的科学资料,建立科学数据库,提供给各方面的研究工作者共同使用。通过科学数据库可以获得不同单位的资料,进行科学研究或实际应用,这称为科学数据共享。科学数据库除了包括各种详细的科学数据外,还收集专家知识(如专家的经验和成果)以及科学工具(如实用的分析软件和图表等资料)。

这里提到的科学数据集成、研究成果集成、专家知识集成、科学工具集成等都是知识资源的集成。如何对有用的科学数据、研究成果、专家知识、科学工具等知识资源进行集成，如何面对用户需求，建立智能型的科学数据库，为社会提供有效的科学数据共享，都是知识集成论需要研究的问题。

理论模型的集成是一种知识集成。模型是对事物的近似描述，在研究一种事物时，人们为了说明这种事物的特性，常提出一些理论模型来对它进行近似的描述。

至于比较复杂的事物，它们具有许多不同的方面。在认识复杂事物的过程中，我们要从不同的角度观察复杂事物的不同方面，提出各种不同的模型，分别从事物的不同方面说明它们在不同条件下的不同特性。

因此，在研究的开始阶段，我们对一些比较复杂的事物可能存在许多不同的理论模型，用来近似描述它们的不同方面。然后由于认识的发展，我们逐步把各种不同的理论模型的长处集成起来，建立起统一的理论模型，这就是理论模型的集成。例如，在原子核理论研究中，早期曾提出过原子核的液滴模型和原子核的壳层模型等，后来发展为原子核的统一模型。

后面章节9.2将谈到对人的选择性注意现象的研究，以及选择性注意的理论模型的集成。

学科体系的集成是一种知识集成。一个大的学科往往包括许多较小的子学科。在学科发展的最初阶段，还没有统一的学科体系。这时分散的子学科呈各自独立的状态，但实际上它们之间存在着联系。在学科发展过程中逐步认识到各种子学科的联系，到条件成熟时，就可以在多种子学科的基础上，用统一的框架建立有系统的、大的学科体系，并且可以进一步开拓新的研究领域，这就是学科体系的集成。后面章节9.3中将讨论心理学学科的结构，以及心理学学科体系的集成。

同一个研究领域中各种科学研究取向的集成是另一种知识集成。在一个研究领域的发展过程中，人们会持不同的观点进行研究，因而出现不同的学术思想和研究思潮，那些影响研究领域发展的研究思潮称为这个领域的研究取向。在一个不成熟的研究领域中常存在着许多不同的研究取向。当这个研究领域发展到一定阶段，相关知识积累达到很丰富的程度，就有条件把各种不同的研究取向的有关内容集成起来，形成统一的研究取向。这是集各种不同研究取向之大成的研究取向。后面章节9.4中将谈到认知科学领域中的各种研究取向，以及认知科学研究取向的集成。

下面着重讨论一种重要的知识集成，即学科交叉。前面第四章已经提到过学科交叉的集成现象。在当代自然科学和工程技术领域中，许多科学技术问题往往涉及多种学科，许多新兴的学科更涉及许多不同学科，如纳米科学技术、生物科学技术、信息科学技术、认知科学技术等。在许多情况下，单独靠一种学科是不能解决问题，只有把几种不同学科的知识和技术集成起来共同研究，才能解决问题。这就提出学科交叉的问题，学科交叉就是不同学科的集成。

以现代天文学为例，它是在学科交叉中发展起来的。传统天文学着重使用光学方法，如光学成像方法及光谱分析方法，观测天体和天文现象，特别是在宇观尺度研究天体形貌、空间分布、天体运动和天体相互作用等方面，取得了丰富的成果。现代天文学除了进行光学观测外，还测量空间的原子核和粒子。在理论方面，我们不但从宇观层面研究天文现象，还从微观层面研究天文现象以及天体内部的过程和天体的演化等。原子核物理学、粒子物理学和传统天文学相结合，产生了原子核天体物理学、粒子天体物理学等交叉学科。

再以神经科学和其他学科的交叉为例。许多学科的科学问题涉及人的脑和心智的活动以及人和环境之间的相互作用，它们和神经科学有密切的联系。现代神经科学的发展使这些学科增加了许多新的研究内容，研究工作出现了新的局面。神经科学和这些学科相结合，产生了大量的交叉学科。

大家熟知的有神经化学、神经-内分泌-免疫学、神经心理学、认知神经科学等。神经化学是神经科学和化学相结合的交叉学科，神经-内分泌-免疫学是神经科学和内分泌学、免疫学相结合的交叉学科，神经心理学是神经科学和心理学相结合的交叉学科，认知神经科学是神经科学和认知科学相结合的交叉学科。

近年来，许多新的交叉学科不断出现，如神经信息学、神经语言学、神经经济学、神经管理学、教育神经科学等。

以神经信息学为例，这是神经科学和信息科学相结合的交叉学科。神经科学的成果启发和促进信息科学的发展，信息科学的成果为神经科学的研究提供新概念和新的研究工具。神经信息学用信息和信息处理的观点研究神经系统的信息问题，如神经信息的载体形式以及神经信息的产生、编码、存储、提取、传输、加工等特性，这些研究又为机器智能提供启发。这门学科还利用现代信息工具和技术，建立不同层次的神经科学数据库，为神经科学研究工作提供数据共享的平台，以便对大量的神经科学数据进行分析、建模和理论研究。

在神经科学和信息科学的交叉研究中，除了这两门学科本身外，还有生物学、医学、数学、物理学、化学、计算机科学和其他工程技术科学等多门学科的参与，其中包括各门学科之间大量的知识集成和技术集成。

又如神经语言学，它是神经科学与语言学相结合的交叉学科：神经科学研究语言现象的神经基础，推动语言学的发展；语言现象为神经科学提供启发以及新的课题。

文理交融是文理相关知识的集成。现代社会发展提出了大量的复杂的科学技术问题，诸如环境、能源、气候、人口与健康、矿藏开发、安全生产、生物多样性保护、经济的可持续发展等，它们既是自然科学和工程技术问题，又关系到各种社会因素，并且产生广泛的社会影响。需要自然科学、工程技术和人文社会科学共同对它们进行研究，这就提出了自然科学、工程技术和人文社会科学交叉的问题。

以现代经济学为例，它是在学科交叉中发展的。过去经济学对经济现象较多进行定性的或半定量的分析和研究。在数理科学提供越来越多的研究工具的基础上，经济学与数理科学相结合，对经济现象的定量研究与定性研究结合起来，产生了数理经济学等交叉学科。

再以神经经济学和神经管理学为例。神经经济学是神经科学与经济学相结合的交叉学科：神经科学研究经济活动中认知过程的神经基础，扩大了对社会认知机制的理解；面向经济活动的神经科学研究成果又推动了经济学的发展。神经管理学是神经科学与管理科学相结合的交叉学科：神经科学有关管理工作脑机制的研究丰富了神经科学的知识，也使管理科学得到新的科学基础。

复杂的科学技术问题需要不同学科共同进行研究，从不同的角度，根据不同的经验，采用不同的方法来研究同一个问题。不同学科的知识和技术是可以互相补充的，通过学科交叉把这些知识和技术集成起来，解决共同的问题。

学科交叉应当是实质性的交叉，而不是形式上的交叉，不是把几种学科简单地合并在一起。为了解决共同的科学问题，这里需要概念和方法的融合。除了知识和技术的集成之外，我们还要组织不同学科的人力和资源，打成一片，步调一致地工作。如果把几种学科放在一起，把一个课题分成关联不多的几个部分，各个学科的人员仍各做各的工作，然后把结果汇总一下，那么是不能解决问题的。这种做法不是真正的学科交叉。我们需要总结学科交叉的经验，以便更有效地进行多学科的知识集成。

学科交叉是长时间的知识集成过程。知识需要不断积累，而不是一朝一夕完成的。因此一个复杂的科学问题的交叉研究，常分成若干阶段，循序渐进。知识集成也是一个探索过程，在长期的实践中不断增加知识、总结经验、修正错误，才能得到新的成果。知识集成过程不但把现有的各方面知识组织起来，而且通过集成，产生新的概念和问题，提出新的假设和计划，再经过新的实验检验，发展原有的认识。这就是知识创新。

9.2 选择性注意的集成模型

这一节介绍选择性注意的集成模型。作为理论模型集成的一个例子，选择性注意是脑的重要功能之一。在有意识的心智活动中，选择性注意起重要的作用，所有各种有意识活动都需要选择性注意参与。

科尔贝塔（Corbetta）等人（1991）曾用正电子发射断层扫描脑成像技术进行选择性注意的实验。实验中，研究者给被试呈现视觉图像，要求被试看图像并报告图像的形状、颜色和运动速度变化。当被试被动知觉刺激时，有一些专一性的脑区激活。而当被试注意刺激时，相同脑区的激活增强；在注意时还有基底神经节和前扣带回脑区的激活。希利亚德（Hillyard）和皮克顿（Picton）（1987）曾用事件相关电位技术进行选择性注意的实验，表明被试在注意时事件相关电位信号明显增强。邓肯（Duncan）、沃德（Ward）和夏皮罗（Shapiro）（1994）在实验中观测到，被试在注意一个项目后数百毫秒内，很难对继续呈现的另一个项目做出反应。选择性注意的一个特点是，在同一时刻只有一个项目能被注意。

为了解释选择性注意的实验事实，我们提出选择性注意的集成模型，即选择性注意的甄别和符合模型，认为选择性注意的机制像是具有甄别和符合功能的电子线路（唐孝威，郭爱克，2000）。甄别电路和符合电路是核电子学中常用的标准的电路，它们常被制成模块，由大量模块构成复杂的电子学系统。

第四章介绍过甄别电路，它们的作用是选择记录输入信号的幅度在一定水平以上的事件，甄别电路对输入信号幅度的限制水平称为甄别水平。第五章介绍过符合线路，它们的作用是选择记录同时输入信号的事件。符合电路可以是二路输入信号符合的电路，称为二重符合电路，也可以是多路输入信号符合的电路，称为多重符合电路。在符合电路同时输入的信号中，一路输入信号可以是设置的控制信号，称为门控信号。

在我们的模型中，不同项目的信号是电子线路的输入信号，它们分别输入到各个甄别电路，再输出到符合电路；注意控制信号是符合电路的一路输入，起门控信号的作用；由甄别电路和符合电路组成的联合电路的作用，是选取输入信号中幅度在一定水平以上且与注意控制信号同时发生的事件。

按照这个模型，许多输入信号在电子线路的许多平行的输入通道中加工。每一个通道中都有一个甄别电路，如果加工信号的幅度超过了甄别水平，那么那个通道就有输出信号。模型要求许多平行的甄别电路的甄别水平会自动调整，使得输入到许多个甄别电路的信号中只有一个通道的信号能够超过甄别水平。这样，那个通道的甄别电路就有输出信号，而在其他通道中加工的信号因为幅度低于统一的甄别水平，所以没有输出。

然后，那个有信号输出的通道，其输出信号到二重符合电路，它是符合电路的一路输入，而二重符合电路的另一路输入信号是注意控制信号。在二重符合电路中，只有当两个信号在符合电路分辨时间之内输入时，才会有符合输出；如果两个信号在符合电路分辨时间之外输入，那么就没有符合输出。所以在符合电路分辨时间之内与注意控制信号同时输入的那个甄别电路的输出信号，会在符合电路中被增强而且记录下来（唐孝威，郭爱克，2000）。

前人曾经提出过许多关于选择性注意机制的模型，例如哈伯兰特（Haberlandt）（1997）的选择性注意的过滤器模型，克里克（1984）的选择性注意的探照灯模型，德西蒙（Desimone）和邓肯（1995）的选择性注意的偏置竞争模型等。按照选择性注意的过滤器模型，选择性注意的机制像是过滤器。按照选择性注意的探照灯模型，选择性注意的机制像是一个心智的探照灯，使被照亮的项目的加工过程得到增强。按照选择性注意的偏置竞争模型，选择性注意的机制像是各个被加工项目之间的竞争。例如，在视觉注意时，视野中的物体会在表征、分析或控制方面进行竞争。这些模型能描述选择性注意的不同方面，也各有不足之处。

选择性注意的甄别和符合模型实际上是选择性注意的集成模型。它把选择性注意的各种特点集成起来，同时又把以上几种模型，特别是探照灯模型和偏置竞争模型集成起来，形成选择性注意的集成模型。

按照选择性注意的集成模型，因为不同项目的输入具有不同强度以及自上而下影响的不同，它们之间会发生竞争。其中有一个项目在竞争中取得优势，原因是它的强度在不同项目中是最大的，引起脑区的激活水平最高。注意的作用是使被注意的项目增强，它并不对每个项目逐个进行扫描，而仅仅使在竞争中占优势的那一个项目得到增强，同时使不受注意的其他项目受到抑制。

选择性注意的集成模型和现有的其他模型的区别是：在探照灯模型中被注意的项目是被动地增强的，但是在选择性注意的集成模型中各个输入项目间主动竞争，只有在竞争中具有最大强度而占优势的一个项目得到增强，而其他项目则被抑制；偏置竞争模型指出了项目之间存在竞争，但是没有强调注意增强作用的重要性，然而按照选择性注意的集成模型，对于选择性注意来说，竞争和增强都是重要的（唐孝威，郭爱克，2000）。

9.3 心理学学科体系的集成

心理学发展到今天，呈现出十分繁荣的局面。在心理学发展过程中，除对心理学自身的各个领域进行深入的研究外，心理学还和许多其他学科交叉研究，产生大量的交叉学科。

目前心理学有很多分支学科，下面列出其中一部分例子：认知心理学、思维心理学、情绪心理学、神经心理学、生理心理学、生物心理学、健康心理学、医学心理学、临床心理学、进化心理学、发展心理学、教育心理学、学校心理学、体育心理学、人格心理学、心理物理学、工程心理学、环境心理学、社会心理学、工业及组织心理学、职业心理学、管理心理学、广告心理学、司法心理学、军事心理学，等等。

有些科学家讨论过心理学的范式问题。按照库恩（Kuhn）（1970）的说法，一门具体学科的科学范式是关于这门学科内容的、全世界几乎公认的世界观，即观察世界的一种方式。他认为科学范式是一门学科的观念框架，它和这门学科的理论和方法，以及普遍接受的一系列科研规则结合在一起。希尔加德（Hilgard）（1987）认为，因为心理学中至今没有一个理论和方法占优势，所以当前的心理学还处于"前范式"的阶段。

面对心理学中各种分支学科林立的局面，心理学需要有一个统一的学科体系。施塔茨（Staats）（1999）、斯滕伯格和格里戈连科（Grigorenko）（2001）、登马克（Denmark）和克劳斯（Krauss）（2005）、斯滕伯格（2005）等曾先后提出对心理学统一理论的期望。一个统一的学科体系将有利于心理学的进一步发展。

在《统一框架下的心理学与认知理论》（唐孝威，2007）一书中，我们以心理相互作用的大统一理论为基础，构建大统一心理学的理论框架，大统一心理学理论也是心理学学科体系的集成理论。从心理相互作用的观点来看，目前心理学中的不同分支学科领域，实际上是研究不同的心理相互作用。以生理心理学和心理生理学两种学科为例，它们都研究心脑相互作用和心身相互作用。生理心理学着重研究脑的生理活动和身体其他生理活动对心理过程的影响，而心理生理学则着重研究心理过程对脑的生理功能和身体其他生理功能的影响。

因为各种不同的心理相互作用具有统一性,所以心理学的各个研究领域有统一性。大统一心理学面对心理现象的全局,可以涵盖心理学的所有研究领域。以心理相互作用的大统一理论为基础的大统一心理学理论,按照各种心理相互作用统一性的原理,考察心理学各个研究领域之间的联系,对它们进行集成的研究,同时发展心理学各个领域间的交叉研究,以及多方面的实际应用。

大统一心理学理论的主要内容包括:心理现象中心理相互作用和各种心理相互作用大统一的观点、心理学具有统一性的观点,以及心理学学科体系集成的观点。这个理论之所以被称为大统一心理学理论,是因为它是在心理相互作用大统一理论的基础上建构的。其中"大统一"是指各种不同种类的心理相互作用的大统一;另一个意思是,在这个理论指导下,把心理学各个研究领域和分支学科集成起来,形成心理学的一个大统一的学科体系。

《统一框架下的心理学与认知理论》(唐孝威,2007)一书讨论了对当代心理学各种分支学科的集成,从概念和方法上把当代心理学的各种分支学科集成为大统一的学科体系。在当代心理学的许多分支学科中,有的分支学科侧重研究心理活动的基础原理,有的分支学科是在心理学和其他一些学科的交叉研究中发展起来的,它们涉及心理学和其他学科共同关心的问题,还有的分支学科根据实际的需要,侧重研究心理学在各种有关领域中的实际应用。

各种心理相互作用的统一性,使得当代心理学中内容千差万别的各种分支学科之间有内在的联系,因而大统一心理学的学科体系能够把这些不同的分支学科集成起来。下面根据心理活动各种成分之间的相互作用、心脑相互作用、心身相互作用、心物相互作用和心理-社会相互作用等不同种类的心理相互作用，对心理学的各种分支学科进行讨论和分类。这些分支学科可以按它们涉及的心理相互作用而分为四类。

心理学分支学科的第一类是与基本的心理特征和心理过程有关的分支学科。这些分支学科研究的问题所涉及的心理相互作用，主要是心理活动各种成分之间的相互作用和心脑相互作用。这一类分支学科有神经心理学、认知心理学、思维心理学、情绪心理学、人格心理学等许多基础学科。

心理学分支学科的第二类是与身体的许多方面有关的分支学科。这些分支学科研究的问题所涉及的心理相互作用，主要是心身相互作用，也涉及心脑相互作用等。这一类分支学科有生物心理学、生理心理学、心理生理学等基础学科，以及与心身关系有关的健康心理学、医学心理学、临床心理学、康复心理学等许多应用学科。

心理学分支学科的第三类是与环境的许多方面有关的分支学科。这些分支学科研究的问题所涉及的心理相互作用，主要是心物相互作用，同时也涉及心脑相互作用和心身相互作用等。这一类分支学科有心理物理学、进化心理学等基础学科，以及与心物关系有关的环境心理学、工程心理学等应用学科。

心理学分支学科的第四类是与社会实践有关的分支学科。这些分支学科研究的问题所涉及的心理相互作用，主要是心理-社会相互作用，同时也涉及心身相互作用和心物相互作用等。这一类分支学科有社会心理学、发展心理学等基础学科，以及与心理-社会关系有关的教育心理学、学校心理学、工业及组织心理学、职业心理学、管理心理学、广告心理学、司法心理学、军事心理学等许多应用学科。

心理学学科体系就是由这四类学科集成的。在这个集成的学科体系中，各种分支学科按涉及的不同的心理相互作用，分别有各自的位置，又按心理相互作用的统一性而有互相间的关联。它们不是杂乱无章的，也不是独立无关的。

包含心理学学科体系在内的大统一心理学理论是基于心理学长期发展的成就并且把它们集成起来的理论，它不和现有的心理学相抵触，而是集心理学众多成就之大成。大统一心理学并不是心理学的一门新的分支学科，也不能代替心理学各个专门领域的具体研究。大统一心理学提出了心理学的一种研究取向，即心理学的统一研究取向。大统一心理学不是终结现有心理学的发展，而是通过大统一和集大成的观念，来开拓心理学进一步发展的道路，促进心理学的进一步繁荣（唐孝威，2007）。

前面提到心理学的范式问题，大统一心理学理论或许可以作为建立心理学范式的候选理论之一。大统一心理学中关于多种心理相互作用的观念和这些心理相互作用统一性的观念，以及关于心理学中不同分支学科集成为统一的学科体系的观念，可能为心理学提供一种观察心理世界和心理学学科的方式。它提供了一个观念框架，在这个框架中有可能把当前心理学中不同的理论和方法统一起来。

它也提供了一种理论，有可能把各种分支学科集成为统一的心理学学科体系。

由于大统一心理学所研究的是所有各种心理相互作用以及它们的统一，它可以包含所有各种研究不同心理相互作用的理论和方法。这样，或许可以做到库恩（1970）所要求的：不同的理论在同一个范式内存在，这些不同的理论各自解决不同的问题，而范式则提供了学科的观念框架。

9.4 认知科学理论的集成

人的认知是非常复杂的过程，认知科学是研究认知现象和规律及其应用的学科。认知科学研究是当代科学研究的前沿之一。

在认知研究中有许多不同的思潮，这些思潮成为认知的不同研究取向。加德纳（1985）和哈伯兰特（1997）等指出，在当代认知研究中有许多不同的研究取向，如神经生物学的研究取向、信息加工的研究取向、具身认知的研究取向、情境认知的研究取向以及社会认知的研究取向等；此外，还有进化心理学的研究取向、发展心理学的研究取向、人工智能的研究取向，等等。

这么多的研究取向之间有什么关系？能不能把它们集成起来？这是一个值得研究的重要问题。在认知科学的统一理论方面，纽厄尔（Newell）在1990年曾经进行过讨论。他认为，科学的目标是统一，心理学需要统一的理论，各种不同的认知过程也需要有统一的理论。在列举认知的各种过程（如问题解决、决策、学习、记忆、技能、知觉、动作、语言、动机、情绪以及想象、梦、白日梦等）之后，他说，需要有把这些不同的认知过程统一起来的理论，即认知的统一理论，并提出了一种可以作为认知统一理论的 SOAR 理论。此外，还有学者曾经建议过别的认知统一理论，如 ACT-R 理论（Anderson，1983；Anderson et al.，2004）。但是斯滕伯格（2005）指出，由于认知的复杂性，自纽厄尔的专著发表以后，至今在认知的统一理论方面还没有满意的方案。

《统一框架下的心理学与认知理论》（唐孝威，2007）一书中对当代认知科学各种不同研究取向有简短的说明。下面引用这些说明：

> 神经生物学的研究取向用神经生物学的观点研究认知过程，着重讨论认知过程的神经生物学基础。这种研究取向认为，人的认知过程和脑内神经活动有密切关系，因此需要了解各种认知过程的不同的神经相关物。这种研究取向关心的问题是：不同的认知过程分别是由哪些脑区参与的，以及认知过程中的神经活动等（Kosslyn & Koenig，1995；Gazzaniga，2000）。
>
> 信息加工的研究取向用脑内信息加工的观点研究认知过程。这种研究取向认为，人的认知过程是脑对环境输入的信息进行编码、存储、提取和操作的过程。这种研究取向关心的问题是认知过程中脑内信息加工的方式和机制（Newell &

Simon, 1972; Haberlandt, 1997)。

具身认知的研究取向用心身关系的观点研究认知过程。具身认知的意思是：认知植根于人的身体，体现于人的身体。这种研究取向认为，认知过程是身体参与的，认知依赖于身体，和身体密切联系而不能分开，因此，认知过程是具身的认知。这种研究取向强调身体影响认知过程，关心的问题是认知和身体的关系，以及身体因素对认知过程的影响等（Varela, Thompson, & Rosch, 1991）。

情境认知的研究取向用认知与情境相关的观点研究认知过程。情境认知的意思是：认知过程是人置身于实际环境时进行的，认知过程依赖于现实情境。这种研究取向强调，认知过程必须置身于现场情境，而不能把两者分开。因此，认知过程是情境的认知，这种研究取向关心的问题是认知过程和现场情境之间的关系等（Gibson, 1979; Brooks, 1991）。

社会认知的研究取向用社会环境作用的观点研究认知过程。社会心理学认为，人是社会的人，社会环境和人的认知过程有密切的关系，因此要研究社会环境与个体认知过程之间的相互作用，研究社会与文化对个体认知过程的影响。这种研究取向关心的问题有家庭、团体、社会对认知过程的作用，以及在社会现场情境中的认知过程等（Cacioppo et al., 2002）。

进化心理学的研究取向用生物进化的观点研究认知过

程。这种研究取向认为，生物学因素如遗传因素对个体的认知过程有重要的作用，强调要研究生物进化与认知过程的关系，以及遗传因素和个体认知过程的关系等。

发展心理学的研究取向用个体发展的观点研究认知过程。这种研究取向认为，在个体一生中认知能力都在变化，强调要研究个体认知的发展过程，包括儿童认知发展的过程，以及童年期认知和成年期认知的关系等。

人工智能的研究取向是用人脑与计算机对比的观点及人工智能的观点研究认知过程，强调要研究类脑的机器以及机器认知等。

由此可见，当代认知的科学研究呈现许多不同的研究取向并存的局面。格拉斯曼（Glassman）（2000）在讨论心理学的基本问题时说："心理学的基本问题之一是如何对付存在着不同研究取向的局面。"如何对付认知科学中存在着不同研究取向的局面，也是认知科学的基本问题之一。面对认知科学研究取向方面众说纷纭的局面，对不同研究取向进行集成，从而建立认知研究的统一理论体系的问题已经提上了日程，也就是说，需要一种认知科学理论，能从观念和方法上把这些不同的研究取向集成起来。

当代认知科学的各种不同研究取向的研究侧重点有所不同，它们都有一定的实验依据和研究特点，同时各有其不足之处。从认知过程涉及的各种心理相互作用来看，它们着重讨论的分别是各种不同的心理相互作用。有些研究取向的不足之处在于，它们只分别研究某一种或某几种心理相互作用，或者只着重于某种心理相互作用的某些方面，而不能涵盖认知过程中各种心理相互作用。但是各种不同研究取向的一些概念并不都是互相排斥的，而是可以互相补充的。各种研究取向有积极的、有益的观点，分别适合于讨论不同的心理相互作用。因此，我们要发挥这些积极的、有益的观点，并且把它们在统一的研究取向中集成起来。既然各种研究取向可以互相补充和互相融合，统一的研究取向就要集它们的积极的、有益的观点之大成。

神经生物学的研究取向关心认知过程的神经生理学基础。神经生物学过程是认知活动的物质基础，为了了解认知的本质，研究认知活动相关的脑功能活动和身体的生理活动是十分重要的。从心理相互作用来说，这种研究取向着重在神经生物学方面研究认知过程中的心脑相互作用和心身相互作用。这种研究取向的许多工作对于了解认知过程中的心脑相互作用和心身相互作用是有益的。其不足之处是很少讨论认知过程中心理活动各种成分之间的相互作用，也不讨论认知过程中的心物相互作用和心理-社会相互作用。

信息加工的研究取向考察认知过程中脑内的信息加工。在认知活动时，脑内有信息加工和意识活动的耦联，用信息加工和意识活动的观点研究认知过程，对于了解心理活动各种成分之间的关系和心脑关系是有益的。此外，研究物理因素对内部心理活动及其脑机制的影响也是令人感兴趣的。从心理相互作用来说，这种研究取向着重研究认知过程中心理活动各种成分之间的相互作用、心脑相互作用和心物相互作用。它的特点是涉及多种心理相互作用中内部信息加工的问题，其不足之处是很少讨论认知过程中心身相互作用，也几乎不考虑认知过程的心身相互作用中有关身体的生理过程，以及身体的生理过程对认知活动的作用。

具身认知的研究取向考察认知过程中身体的因素，情境认知的研究取向考察认知过程中环境的因素。这些对认知研究来说都是重要的。具身认知的研究取向强调研究认知与身体的关系，从心理相互作用来说，这种研究取向侧重讨论认知过程中的心身相互作用。情境认知的研究取向强调研究认知与情境的关系，从心理相互作用来说，这种研究取向侧重讨论认知过程中心理活动与环境间的相互作用。这两种研究取向的不足之处是，它们很少讨论认知过程中心理活动各种成分之间的相互作用和心脑相互作用。

社会心理学的研究取向关心个体心理和社会环境之间的关系。社会环境和文化对认知过程有很大影响。这种研究取向的观点对于了解认知与社会的关系是十分重要的。从心理相互作用来说，这种研究取向主要涉及认知过程中的心理-社会相互作用，还讨论社会环境对认知过程中心理活动各种成分之间的相互作用以及心身相互作用的影响，其不足之处是很少讨论认知过程中的心脑相互作用和心物相互作用。

认知科学的其他一些研究取向，如进化心理学的研究取向、发展心理学的研究取向和人工智能的研究取向等，也各有特点。例如，进化心理学的研究取向关心生物进化对心理和行为的作用。从心理相互作用来说，这种研究取向涉及认知活动中多种心理相互作用。

用这种研究取向的观点研究认知活动，对于了解进化过程对认知活动中的各种心理相互作用有哪些影响是有益的。又如，发展心理学的研究取向关心人生全过程，特别是婴幼儿、童年、青少年时期心理和行为的发展。从心理相互作用来说，这种研究取向涉及认知活动中多种心理相互作用。用这种研究取向的观点研究认知活动，对于了解认知活动各种心理相互作用发展变化的特点是有益的。

认知科学的集成理论用各种心理相互作用及其统一的观点考察当代认知科学的各种不同研究取向，并且把这些研究取向中有益的观点集成起来。这种集成的研究取向也被称为认知科学的统一取向。下面说明这种研究取向关心的问题、考察的重点，以及研究的观点和方法。

认知科学的统一研究取向关心认知过程中所有各种心理相互作用以及它们的统一性。这种研究取向要求对认知过程涉及的所有各种心理相互作用进行全面的考察和研究，还指出不同的心理相互作用具有共同的基础，因而要对认知过程中所有心理相互作用进行统一的研究。这种研究取向不但分析认知过程中每种心理相互作用的特点，而且要考察这些心理相互作用变化的动态过程。

如前所述，当代认知科学的各种研究取向分别侧重于讨论认知过程中不同种类的心理相互作用，而认知科学的统一研究取向则讨论认知过程中所有各种心理相互作用以及它们之间的联系和统一。因为它的研究内容全面地包括所有各种心理相互作用，所以这种研究取向可以包容认知科学的各种不同的研究取向，把它们的积极的、有益的内容集成起来。总之，认知科学的统一研究取向和以往其他研究取向的不同之处在于，它讨论的不仅是认知过程中某几种心理相互作用，或某种心理相互作用的某些方面，而是所有各种心理相互作用，以及各种心理相互作用的所有方面；不仅是认知活动中各种心理相互作用的特性，而且是认知活动中所有各种心理相互作用的动态过程。

认知科学的统一研究取向先对当代认知科学各种研究取向进行详细的分析，分别研究它们的特点，提取它们的有益内容和方法，然后在取其精华的基础上，把它们在统一的理论框架中集成。因此它的研究领域包含认知科学的全部领域，比目前其他的研究取向关心的领域更加广泛。

认知科学的统一研究取向认为，当代各种认知研究取向分别讨论的问题都是认知活动的重要方面。认知过程既有神经生物学的基础，又有信息加工与意识活动的特点；认知过程既是具身的，又是情境的和社会的；认知既是进化的，又是发展的。认知过程涉及所有各种心理相互作用，它们之间又有紧密的联系，在认知过程中有各种心理相互作用集成的过程。

第十章　工程集成论

在工程技术领域中存在各种集成现象。将一般集成论的观点应用于工程技术领域，研究这些领域中集成现象的特点，对解决工程技术中的复杂问题是有帮助的。

在一般集成论的应用中，我们提出建立一门研究工程领域集成现象和规律及其应用的学科，并把它命名为工程集成论，它的英文名称是 engineering integratics。我们又提出建立一门研究技术领域集成现象和规律及其应用的学科，并把它命名为技术集成论，它的英文名称是 technology integratics。

工程集成论和技术集成论的研究内容非常丰富。前面第四章提到过技术领域的集成现象。这一章举出工程集成和技术集成的几个例子进行讨论，它们是大科学计划的集成、大型实验装置的集成和医学影像技术的集成。

10.1 大科学计划的集成

在现代科学技术的发展中出现了一类规模巨大的科研项目,由于这些项目要解决的科学技术问题非常庞大复杂,对这些项目需要制订长期的计划,投入大量的资金,组织许多科研单位和科研人员共同实施。相对于在一个实验室里由一个科研团队进行的规模较小的科研项目来说,它们的规模很大,因此被称为大科学计划或大科学项目。大科学计划是现代科学技术研究的一种研究方式,它们具有许多不同于规模较小的科研项目的特点。

美国在科学技术方面实施过的几个计划,如曼哈顿计划、阿波罗计划和人类基因组计划等,都是大科学计划的例子。曼哈顿计划是20世纪40年代美国研制原子弹的工程,阿波罗计划是20世纪六七十年代美国人登月的工程,人类基因组计划是20世纪90年代对人类基因组测序的工程。目前世界各国联合进行的国际热核聚变实验堆计划也是大科学计划的例子,这个计划是研制热核聚变反应堆装置的工程。我国也成功地实施过几个大科学计划,并且正在实施几个新的大科学计划。

兰布赖特(Lambright)(2009)在《重大科学计划实施的关键:管理与协调》一书中论述了大科学计划的实施,介绍了美国几个大科学计划的案例。书中详细叙述了人类基因组计划的发展历程和项目管理的经验,还提到在这个计划实施过程中一些有趣的故事。这本书从目标、组织、支持、竞争、领导等方面介绍大科学计划的管理工作,指出管理与协调是实施大科学计划的关键。书中还考察了另外几个大规模的研究和开发计划的案例,包括纳米技术计划和国际空间站计划。

从一般集成论的观点来看,在大科学计划的规划和实施过程中存在许多集成现象。大科学计划既要解决科学技术问题,本身又是工程项目。大科学计划中的集成问题是工程集成论研究的重要课题之一。

从大科学计划的酝酿阶段开始,到工程的规划、立项、实施直至完成,在大科学计划的每个阶段中都有许多不同形式的集成过程。兰布赖特提到的管理与协调,是大科学计划中集成现象的重要内容。大科学计划的组织和管理都是集成的过程,如各种科学思想和科研方案的集成、多种学科和多种技术的集成、不同科研单位和科研团队的集成等。在工程的集成过程中要进行大量的协调工作,使大科学计划得以高效率、高质量的完成。

实际上,不仅在大科学计划的规划和实施中有许多集成过程,在所有科研工作中,也都有组织、管理、协调等集成过程,只是在大科学计划中集成过程具有更大的规模,有更加显著的表现,并且起更加重要的作用。科研集成论需要研究这些集成过程的特点和规律,并且按照这些特点和规律来指导科学计划的实施。

10.2 大型实验装置的集成

大型实验装置的设计、建设和运行中的集成是工程集成和技术集成的一个例子。现代科学实验除了在许多实验室中进行各类小型实验之外，还建立起一些大型科学实验中心，其中装备了许多大型实验装置，这些大型实验装置可以提供给许多用户单位共同使用，进行多种科学实验。

大型实验装置是复杂的工程设施，它包括许多工程系统，由这些配套的工程系统集成为整体的大型实验装置；其中每个工程系统又包括许多实验设备，由这些配套的实验设备集成为功能完整的实验系统。

例如，同步辐射光源实验中心的大型实验装置有产生同步辐射光的装置和许多实验站等实验系统。产生同步辐射光的装置包括把电子加速到高能量的加速器和使高能电子稳定运转的环形装置。第五章曾经提到过同步辐射，高能电子束在环形轨道上做圆周运动时产生同步辐射光，由许多与圆周成切线的管道引出同步辐射光，分别送到许多实验站。每个实验站配置各种靶和专门的探测设备。例如，用电子能谱仪测量同步辐射光打靶后反应产物的特性。

又如，高能物理实验中心的大型实验装置，有各类加速器系统和各类探测器系统。其中加速器系统加速带电粒子，提供高能物理实验所需的粒子束；探测器系统有许多专用设备，测量高能物理反应产物的特性。

下面以欧洲核子研究中心的大型正负电子对撞机（LEP）为例子，来说明这种大型实验装置的复杂性和集成性（唐孝威，1985；1993）。正负电子在对撞机中实现对撞，对撞时正负电子的总能量达到2000亿电子伏。这个能区可以产生大量的中间玻色子Z^0粒子和W$^±$粒子，因此LEP称为中间玻色子"工厂"。

这台对撞机的主体建于法国和瑞士的交界地区。主环和对撞区的几个实验大厅都建在地下，离地面最浅为50米，最深达170米。主环的周长是27千米。对撞机系统的主要部分有加速电子的装置、产生和加速正电子的装置以及正负电子的对撞区等。正负电子在主环中以相反的方向运动，在主环的四个对撞区实现正负电子对撞。这台对撞机于1983年动工，至1989年建成并投入运行。

围绕着四个对撞区分别安装着四个大型探测器，有OPAL探测器、ALEPH探测器、L3探测器和DELPHI探测器，分别测量高能正负电子对撞的产物。每台探测器系统都是庞大的实验装置。以其中的L3探测器为例，它是一个覆盖4π立体角、内有大体积磁场的大型探测器。这个探测器的特点是以很高的精确度测量光子、电子和μ子。探测器总重量达8000吨，在体积为$12×12×14$立方米的空间中，用常规磁铁产生强度为0.5特斯拉的沿束流方向的均匀磁场（Adeva et al., 1990）。

从对撞区朝外看，L3探测器由以下各层组成：（1）围绕着对撞区的是一个顶点探测器，称为时间扩展室，在它外面还有四层测量粒子水平方向坐标的正比室；（2）在顶点探测器外面是由锗酸铋晶体组成的电磁量能器，大量的条状锗酸铋晶体对准对撞区，把对撞区包围起来；电磁量能器测量能量大于20亿电子伏的光子、电子的能量分辨率优于1%；根据顶点探测器和电磁量能器给出信息的组合，可以区别光子与电子；（3）外面是强子量能器，围绕着束流线的桶部强子量能器中共有144个量能器模块，还有端盖部强子量能器覆盖两端；它们由铂板和正比室夹层组成，用它们可以测量入射强子的能量和位置；(4) 再外面是大型的μ子漂移室，它们测量能量为500亿电子伏μ子的动量分辨率为2%；此外，在对撞点两侧各2.75米处，放置束流亮度监测器。

研究者用L3探测器得到了许多令人感兴趣的物理成果，比如Z^0粒子和$W^±$粒子特性的测量、中微子代数的确定、电弱相互作用参数的测定、Z^0粒子衰变为b夸克对的衰变特性测量、量子色动力学检验与强耦合常数的确定、量子电动力学检验及轻子特性测量等(L3 Collaboration, 1993; 唐孝威, 1993)。

大型实验装置的集成问题是工程集成论的一个重要研究课题。建造大型实验装置的目的是利用这些装置进行科学实验。实验装置的效益表现在装置指标先进、运转良好、工作稳定，有效利用开机时间，取得大量的、高质量的实验成果。实验装置中各个系统和各种设备的正确而有效的集成以及协调而稳定的运行，对提高装置的整体利用率和实验成果的产出率起重要的作用。

大型实验装置从开始规划和设计到建造和运行，都要考虑优化集成。以高能物理实验中心为例，加速器系统要优化集成，其中各种实验设备和部件（如离子源、电源、磁铁、控制系统、真空系统等）都要长期稳定、可靠、有效地工作，并且相互之间要有良好的接口。各种实验设备配套齐全，协调运行。探测器系统也要优化集成，包括硬件和软件的集成。探测器系统中各种实验设备和部件（如探测装置、分析磁铁、电子学、计算机、屏蔽系统等）都要可靠地工作，并且相互之间要配套齐全，保证实验仪器稳定运行。

从大型实验装置的组织管理工作来说，大型实验装置工程建设的规模很大，需要庞大的工程队伍和实验队伍参加，他们团结一致，协调工作，才能出色地完成任务。因此团队集成也是工程集成的重要部分。

10.3 医学影像技术的集成

医学要对人类疾病进行预防、诊断和治疗。只有正确诊断疾病，才能对疾病进行有效的治疗。常规的医学诊断方法有人体各种生理样品的化验和人体各项生理指标的测量等。科学技术的发展使得医学诊断可以无创伤地"透视"人体，"观测"和"拍摄"的不仅有人体内部结构的图像，而且有人体内部功能的图像。医学影像技术已经成为现代医学诊断的重要手段。

以射线成像为例，利用射线和人体相互作用，我们可以探测人体内部的结构和功能。在结构成像方面，我们可以在体外用射线照射人体，射线对人体有穿透力，在体外测量透过的射线。射线在人体不同部位有不同的穿透，这种对比度提供人体内部结构的信息。在功能成像方面，我们可以将含有微量放射性的示踪剂注入人体，它们参与人体功能活动，在体外测量它们发射的、穿出人体的射线。不同种类的示踪剂在人体内部不同部位参与不同的功能活动，可以提供人体内部功能的信息。在体外测量射线时可以对人体进行逐层扫描，得到断层的图像，再由此重建为立体的图像。这些人体内部的结构图像和功能图像，为临床诊断疾病提供了直接的依据。

目前医学影像技术的种类很多。医院中常用的技术有X射线计算机断层扫描（CT）、磁共振结构成像（MRI）、功能磁共振成像（fMRI）、正电子发射计算机断层成像（PET）、单光子发射计算机断层成像（SPECT）等（唐孝威，1999；2001）。

从人体的结构成像来说，计算机断层扫描和磁共振成像可以无损伤地测量人体内部结构。前者是用X射线（或γ射线）照射人体，通过多次投影来获得人体内部结构的图像。后者是利用磁共振原理测量人体内部质子的分布，来获取人体内部结构的图像。这些影像技术在疾病的诊断和治疗方面发挥了重要作用，但是它们给出的图像是人体的结构图像而不是功能图像。

功能成像不同于结构成像，它能显示人体内发生功能变化的区域及其时空特征。以脑功能成像为例，若要知道人脑高级功能活动在哪些脑区发生以及这些脑区的功能连接，或了解疾病状态时脑功能有什么变化，则要采用脑功能成像技术来研究。近年来在磁共振成像技术的基础上发展起来的功能磁共振成像技术和磁共振波谱技术是进行功能成像的重要手段。

功能磁共振成像的原理是，磁共振信号与血流中含氧量有关，测量脑活动时脑内各处血流中含氧量的变化可反映相应脑区的神经细胞活动的变化。因为当个体执行各种认知任务时，脑局部兴奋，血流增加，但氧耗量的增加小于血流量的增加，血液中脱氧血红蛋白减少。脱氧血红蛋白是顺磁性物质，可使磁共振成像的特征量$T2$延长，$T2$加权像信号增强（磁共振成像有许多特征量，如$T1$和

T2都是弛豫时间，T1是纵向弛豫时间，T2是横向弛豫时间）。这种效应被称为血氧水平依赖性效应，功能磁共振成像的原理正是基于这种效应（Logothetis et al., 2001）。此外，磁共振波谱技术的原理是磁共振信号有化学位移，通过化学位移可以测量体内有关区域中各种化合物分子的谱。

第四章中已经提到过核医学。核医学中的单光子发射断层成像技术和正电子发射断层成像技术是进行脑功能成像的重要技术。前者是把发射γ射线的核素标记的化合物注入人体，使它们进入脑部，在体外测量γ射线而获得这种标记化合物在脑内分布的断层图像。后者是把发射正电子的核素标记的化合物注入人体，使它们进入脑部，在体外测量正电子湮灭发射的γ射线而获得这种标记化合物在脑内分布的断层图像。通常用18F标记的葡萄糖获得脑内代谢的图像，可以进行脑内葡萄糖代谢功能的定量测量；或用15O标记的水获得脑内血流的图像，可用以进行人认知活动中脑激活区的定位。用这些技术得到脑功能的三维图像，其空间分辨率为数毫米。

正电子发射断层成像的特点是：三维（可增加灵敏度和统计精度）；活体（无损，自然生理状态）；动态和功能；定量；正电子核素寿命短，且为生物体组成部分；示踪剂多样性和专一性；等等。但用这种技术得到的脑功能图像空间分辨率较差，须与高分辨率的结构成像结合定位。利用这种技术可以检测脑功能活动的局域能量消耗的变化、神经活性物质在不同脑区分布的特点等，从而了解视觉、听觉、语言、思维等功能活动发生的脑区，也可进行精神分裂症、失语症等疾病的脑内定位。

在脑功能成像方面，除上述技术外，还有其他的成像技术，如基于脑活动时脑内组织的光学性质变化的多种光学成像技术，可以提供观察大脑皮质功能构筑的高分辨率图像。近红外光学成像和光学相干层析成像等技术迅速发展，成为很有前景的新的脑功能成像技术。前者是利用脑内活动对近红外光传输的影响来成像，后者是利用光学相干原理进行脑组织的层析成像。

用医学影像技术获得的图像有人体结构图像和人体各类功能图像。在脑功能图像中有静息态脑功能图像或执行不同任务时的各类脑功能图像。图像的特性包括空间分辨率和时间分辨率，影像装置要给出高分辨率的图像，还要对获得的图像进行准确的处理，以达到更好地诊断和治疗疾病的目的。

在医学影像技术的各个层次和各个方面都存在集成现象。每一种医学影像技术都是相关硬件和软件的集成。在硬件方面有射线源、探测器、电子学、计算机等多个部件的集成，要求各种部件配套，而且性能匹配。各个部件的性能都会影响图像分辨率，提高分辨率要从解决其中的瓶颈问题入手。在软件方面，图像处理时有图像分割、配准、融合等技术的集成。从医学影像装置的运行来说，有数据获取和数据处理的集成。从图像的种类来说，有结构图像、生理功能图像、与任务相关的功能图像的集成。

不同种类的医学影像技术称为不同的模态。例如，CT和MRI是两种模态。由不同模态获得的诊断数据是可以互相补充的。医学影像技术的发展趋势之一是多模态的集成，合理地联合使用多种影像技术，以便更有效地诊断和治疗疾病。这方面的一个例子是

PET/CT装置的建造和运行，这种装置集传统的PET和CT技术为一体，集结构成像和功能成像为一体。另一个例子是与射线影像技术结合的放射治疗装置的建造和运行，这种装置集医学影像技术和放射治疗技术为一体，集医学诊断和医学治疗为一体。

医学影像技术发展的另一个趋势是在分子水平上对人体内部的活动进行成像，这种技术称为分子影像技术。这种技术和传统的结构成像技术及功能成像技术集成起来，不但可以得到宏观的人体内部图像，而且可以得到分子水平上的人体内部活动图像，从而更好地了解疾病的分子机理，有助于疾病的诊断和治疗。分子影像技术的几个要素是分子探针、信号放大、高灵敏探测和分析软件，需要把这几个方面有效地集成起来。

第十一章 教育集成论

在教育领域中有各种集成现象。将一般集成论的观点应用于教育领域，研究教育领域中集成现象的特点，具有实际的意义。

我们提出建立一门研究教育领域集成现象和规律及其应用的学科，并把它命名为教育集成论，它的英文名称是 education integratics。同时在与教育相关领域中，提出建立一门研究人类智能活动中集成现象和规律及其应用的学科，并把它命名为智能集成论，它的英文名称是 intelligence integratics。这一章讨论教育集成论的两个问题：一是智能的集成，二是教育内容的集成。

11.1 智能的集成

智能是人的心智能力和行为能力的集成，智能发展是各种能力的集成过程。教育对智能的发展起重要作用，智能集成有其自身的规律，要根据智能集成的特点进行教育工作。

在《智能论——心智能力和行为能力的集成》(唐孝威，2010)一书中对智能的集成有过详细的论述。下面是该书有关部分的摘要：

> 智能是非常复杂的现象。在智能活动中存在各种各样的集成现象，因此有必要对智能集成现象进行专门的讨论，着重讨论智能活动中的集成作用和集成过程，智能集成作用是智能活动的重要机制，智能集成过程是智能活动的重要内容。
>
> 智能集成论用集成的观点研究智能活动的结构和过程，特别是其中的集成作用和集成过程，从而探讨智能的本质。作为一种理论，智能集成论是关于智能本质的理论，是关于心智能力和行为能力集成规律的理论。作为一门学科，智能集成论是研究智能活动中各个层次和各种类型的集成现象和规律及其应用的学科。
>
> 智能集成论的研究范围是智能现象，主要涉及与智能有关领域的集成现象。下面从研究对象、理论概念、研究内容和研究取向等方面，说明智能集成论的特点。
>
> 从研究对象来看，智能集成论的研究对象不同于以往一些智能理论的研究对象。智能既有心智能力，又有行为能力。在智能活动中，存在许多种心理相互作用，如心理成分相互作用、心脑相互作用、心身相互作用、心物相互作用和心理-

社会相互作用等。以往有些智能理论只关心某些方面的能力，只考察某种或某几种智能成分，只涉及智能活动中某种或某几种心理相互作用；智能集成论则考察所有智能成分，并且涉及智能活动中所有心理相互作用，着重研究它们的集成。

从理论概念来看，智能集成论的理论概念和以往一些智能理论的理论概念不同。智能集成论的核心概念是集成。智能现象涉及不同层次和不同性质的智能成分、集成作用、集成环境、集成过程，以及集成构建的各种智能集成体。不同种类的智能成分，通过它们之间的各种相互作用进行不同形式的集成过程，构成不同层次和不同性质的智能集成体，在一定条件下智能集成体涌现新的特性。以往有些智能理论，如智能的因素理论，分析智能因素但不讨论其关联与发展。智能集成论则强调智能活动涉及的各个层次的智能成分的集成，以及智能活动中各种心理相互作用的集成。

从研究内容来看，智能集成论的研究内容和其他一些智能理论的研究内容不同。智能集成论认为，智能活动不是智能成分的简单叠加，而是通过智能成分间的相互作用进行集成，因此研究内容着重于智能的集成作用和集成过程。智能活动中有多种多样的集成现象，由于不同个体的智能成分、集成作用、集成环境和集成过程具有多样性和复杂性，因而不同个体的智能千差万别。

个体的智能是在先天遗传的基础上，通过各种心理相互作用，在后天长期实践的集成过程中发展的。经验表明，健全的智能成分、能动的集成作用、丰富的集成环境以及协调的集成过程，是有效发展智能、提高智能水平的关键。

从研究取向来看，智能集成论的研究取向和以往许多智能研究取向不同。以往许多智能研究取向以及各种具体的智能理论往往侧重描述智能的某一个侧面或某一些侧面；而智能集成论认为，智能是多种智能成分集成的统一体，智能活动中又有各种心理相互作用的集成，所以在构建理论时要对各种智能研究取向和对各种具体的智能理论进行集成。智能集成论是集各种智能研究取向和各种具体智能理论之大成的理论。

智能集成论的研究包括两个方面：一个方面是研究智能的集成现象，特别是研究智能活动中的集成作用和集成过程；另一个方面是研究智能理论的集成，其中有对当代各种智能研究取向的集成和对各种具体智能理论的集成。研究智能的集成以及研究智能理论的集成，目的都是了解智能的本质和探讨提高智能水平的方法，因此在智能集成论中对智能集成的研究和对智能理论集成的研究这两个方面是一致的。

《智能论——心智能力和行为能力的集成》（唐孝威，2010）一书提出了智能的一个理论框架，其中包括基于心理相互作用及其统一理论的广义的智能定义和智能的统一研究取向、关于智能结构和智能过程的观点，以及智能集成论。

智能集成论是这个智能理论框架的一部分，它用集成的观点考察智能现象和探讨智能的本质。智能集成论包括智能集成的研究和智能理论集成的研究两个方面。

在智能的集成方面，智能集成论的要点如下：

第一，个体智能活动的基础是脑和身体。智能活动包括心智活动、行为活动，以及它们之间的耦联。心智活动是脑的功能。心智支配行为，行为活动由身体实现，个体通过身体的感觉器官、运动器官、语言器官等与外界环境相互作用。

第二，智能活动不能离开环境。在心-脑-身体-自然环境-社会环境的统一体中存在着各种相互作用。个体智能活动是在自然环境和社会环境中进行的，个体所处的多种多样的自然环境和社会环境对智能集成有重要的影响。

第三，智能是心智能力和行为能力的集成。心智能力和行为能力有紧密的联系。心智能力和行为能力结合在一起，构成智能的整体。心智能力和行为能力都是通过不同层次、不同种类的集成过程而集成的，它们都具有复杂的结构。

第四，有不同层次的、多种多样的智能成分和具体能力。心智能力是智能的一个方面，它是由觉醒-注意能力、认知能力、情感能力、意志能力等各种智能成分集成的；其中每一种智能成分又包括许多具体能力，如认知能力包括感觉能力、知觉能力、记忆能力、思维能力、语言能力等许多具体能力；其中每一种具体能力又有结构，如思维能力是由分析能力、综合能力、理解能力、推理能力等集成的。

行为能力是智能的另一个方面，它是由运动能力、操作能力、适应能力、社会能力等各种智能成分集成的；其中每一种智能成分包括许多种具体能力，如社会能力包括人际行为能力、管理行为能力、表达能力等许多具体能力。

智能不是单一的一种成分，而是有许多种成分；不是单一的一种具体能力，而是有许多种具体能力。各种智能成分和具体能力结合在一起，智能是集各种智能成分和具体能力之大成的复杂的统一体。对于复杂的智能，不能只用一种指标来描述。

第五，智能活动中存在不同层次的、多种多样的集成作用。智能活动中有多种心理相互作用，智能活动是通过多种心理相互作用实现的。

从心智活动来说，觉醒-注意、认知、情感、意志等各种成分，通过彼此间的相互作用以及心脑相互作用集成为心智活动。从心智活动中的认知过程来说，感觉、知觉、记忆、思维、语言等各种成分，通过彼此间的相互作用以及心脑相互作用集成为认知活动。从认知活动中的思维过程来说，分析、综合、理解、推理等通过彼此间的相互作用以及心脑相互作用集成为思维活动。从行为活动来说，也有类似的情形。在智能活动中，各种心理相互作用把不同的智能成分集成起来；在各种集成统一体中，这些智能成分不是简单的叠加，而是有机的集成。

第六，在心智活动和行为活动中存在不同层次和不同种类的集成过程。在智能活动的集成过程中，许多不同的智能成分通过集成作用而形成不同层次的各种集成统一体。

心智活动和行为活动都是复杂的过程。在心智活动和行为活动的集成过程中,常常存在优化、同步、协调等现象。

第七,智能集成过程是主动的过程。在心智活动中和行为活动中,存在主动的集成过程。心智和行为通过主动的集成过程形成统一体。以认知活动中的记忆为例,人的长时记忆不是事件的堆砌,而是通过集成过程主动地对记忆资料进行组织,形成既有分类又有联系的记忆网络。

在一定条件下,智能集成过程中会涌现新的功能。例如,在长期的、主动的思维活动中,可能会出现新的思路,从而得到新的结果。

第八,智能的发展性。从种系进化来说,人类的智能是进化的产物。从个体一生的发育和生长来说,个体各种智能成分和具体能力都不是固定不变的,而是在先天遗传的基础上、在后天实践中通过智能集成过程而不断发展的。因此,集各种智能成分之大成的整体智能是不断发展的。

智能集成是一个不断进行的过程。集成过程常具有阶段性,而不是一次完成的。要研究智能的各种成分和具体能力的发展,研究它们随着时间变动的规律。

第九,智能的个体差异。由于个体先天条件有差别,以及个体后天实践和学习过程中智能成分、集成作用、集成环境和集成过程的多样性,不同个体的智能有差异。

不同个体的智能成分和具体能力千差万别,但总是各有所长,又各有所短,对不同个体的智能不能用同一种标准一律要求。

第十,智能的培养和提高。个体的智能是可以培养的,智能水平是可以提高的。许多因素对智能有影响，除遗传和营养等生物学因素以及环境和教育等因素外，自觉的学习和实践对智能水平的提高起决定性的作用。

提高智能水平并不取决于单一因素，而要从许多方面入手，要通过长期的、主动的学习和实践，促进各种智能成分和具体能力的协调发展。

在智能理论的集成方面，智能集成论的要点如下：

第一，智能理论的发展。智能集成论强调智能理论的集成过程，认为智能理论发展的过程是在已有的智能认识的基础上，对智能研究中的新现象、新概念、新理论进行集成的过程。随着人们对智能认识的扩展和深化，智能理论不断发展。

第二，智能研究取向的集成。当代智能研究有多种不同的研究取向，它们分别侧重考察智能活动中不同种类的心理相互作用。基于心理相互作用及其统一理论的智能的统一研究取向，是对当代智能研究的各种不同研究取向进行集成的一种新的研究取向。

智能集成论是根据这种研究取向提出的智能理论，它对智能活动中所有心理相互作用进行全面的研究，而且注重智能活动中各种心理相互作用的统一性。

第三，具体智能理论的集成。智能集成论是对当代各种具体智能理论的有益成果进行集成的理论。现有的多种多样的智能理论各有一定的依据和长处，它们可以互相补充。智能集成论把这些具体智能理论的有益成果集成起来，因而包含了智能的认知理论、智能的因素分析理论、智能的生物学理论、智能的情绪理论、智能的具身理论、智能的情境理论、智能的社会理论等许多理论的有益成果，形成比较完整的智能理论。

虽然智能集成论包含大量的具体智能理论的有关内容，但它并不是这些具体智能理论的简单汇总，而是按照心理相互作用及其统一的观点，对这些具体智能理论的许多有益成果进行集成而构建的新理论。

上面讨论的智能集成的特点对教育工作是有启发的。教育工作应当符合智能集成的特点，有效地促进人的心智能力和行为能力的集成。

11.2 教育内容的集成

人的素质包括德、智、体、美等方面，上一节讨论了智能，它是人的素质的一部分。教育的目标是塑造德、智、体、美全面发展的人。全面的素质教育是对德育、智育、体育、美育四个方面进行集成的教育。

德育是品德教育，要培养人的品德：人人都要热爱祖国，热爱人民，要有远大的理想和高尚的道德，要尊老爱幼，遵纪守法。智育是知识技能教育，要增长人的知识和能力，人人都要热爱科学，热爱学习，要有正确的思想方法，掌握科学知识和技能，培养解决实际问题和为社会服务的能力。体育是体魄教育，要培养强健的体魄，人人锻炼身体，心身健康。美育要让人懂得真、善、美的标准,要有追求美好理想的坚强意志。

人的德、智、体、美是互相联系且统一的。四种素质要全面发展,教育中德育、智育、体育、美育四种内容也互相联系,在教育工作中这四个方面都不能缺少。它们应当并重,而不能偏废。教育的集成强调四种教育内容的集成,通过集成的素质教育来全面地塑造人。人有个体差异,每个人有自己的特点,教育要因材施教,培养富有特色的人;但是德、智、体、美四方面素质的全面发展,对每个人来说都是必要的。要在全面发展的基础上,发扬每个人的特长。

教育神经科学是神经科学和教育科学相结合的交叉学科,这门学科进行面向教育理论和教育实践的神经科学研究。德、智、体、美四方面素质都有相应的神经基础。脑的功能具有统一性,这四方面的神经基础也有统一性。研究德育教育、智育教育、体育教育、美育教育等实践中教育者和受教育者脑活动的机制,将使神经科学的研究内容更加丰富多彩。

这门学科还发展基于神经科学研究成果的教育理论,进行基于神经科学研究成果的教育实践。关于这四方面素质的神经基础的知识和脑功能统一性的知识,是进行教育内容集成的理论基础。把脑科学的研究成果应用于教育实践,将使教育工作能更加有效地提高人的素质。

在神经科学和教育科学的交叉研究中,除这两门学科本身之外,还有生理学、心理学、医学、认知科学、管理科学等多种学科的参与,其中涉及各门学科之间大量的知识集成。

脑的发展有敏感期，在敏感期中进行相应的教育，会有最好的效果。从婴幼儿、儿童、少年到青年，是一生中培养四方面素质的关键阶段。素质教育要从婴幼儿抓起，使他们的身心从小就健康成长。从儿童、青少年到成人，各级学校要创造一个德、智、体、美四方面素质并重的环境，对学生进行全面的素质教育。

但教育不限于儿童和青少年，中年人和老年人也要不断提高素质。人的素质教育是终生的，社会对中年和老年人也要进行素质教育，老年人要学到老，对社会贡献到老。素质教育不仅是学校教育的任务，还是全社会的任务。社会环境是塑造人的大环境，要在全社会营造一个有利于人人学习、人人素质全面发展的良好风气。

在教育工作中如何更好地把德育、智育、体育、美育四个方面的教育内容集成起来，是一个需要深入研究的课题。近年来一些教育家提倡"做中学"的教育方法，就是在动手做的过程中进行科学教育，已经取得很好的效果（韦钰，Rowell，2005）。从全面培养德、智、体、美四方面素质的角度，我们可以把"做中学"的内容加以扩充，在动手做中不但可以学习各种科学知识，还可以提高品德、锻炼身体和培养情感。

附录

谢林顿关于神经系统集成作用的部分论述

谢林顿在1906年发表《神经系统的整合作用》一书,强调研究神经系统整合作用(即集成作用)的重要性。

他在讨论神经生理学时说,可以用三个观点研究神经生理学:

一是神经营养的观点。活的神经细胞和其他活细胞一样,都有生理活动,所以要考察活的神经细胞和神经系统的营养问题。

二是神经传导的观点。神经细胞能够传导神经冲动,所以要考察神经系统的传导过程。

三是神经系统整合作用的观点。多细胞动物的神经系统将动物体内各个器官联系起来。

在这三个观点中,特别重要的是神经整合的观点。

他认为动物体内存在多种整合:一是从结构方面看的动物器官的整合,以及由单个细胞组成统一的动物整体;二是动物体内的化学整合,如体内各种腺体的协调活动;三是血液循环的整合作用,通过血液循环实现机体的统一活动;四是神经系统的整合作用。

他指出，神经系统的整合作用与其他几种整合不同。神经系统的整合作用不是通过细胞间物质的输运来实现，而是通过神经信号的传导来实现的，因此在时间上是高速进行的，而且可以有远程的传导。神经连接有精确的空间分布，所以神经传导有精确的时间分布。

他对中枢神经系统的活动进行了系统的研究。在《神经系统的整合作用》一书中，他提出了"中枢神经系统的作用在于整合作用"的著名论断。他研究过神经系统不同层次的整合作用。

他从研究脊髓反射入手，通过肌紧张反射和屈反射，对中枢神经系统的整合作用进行了深入的研究。他认为反射是中枢神经系统的基本活动方式之一。身体的感受器接受外界刺激，转变为神经冲动，由传入神经传输到中枢，中枢再将神经冲动通过传出神经传输到周围器官的效应器，引起它们的活动，这就是反射。

因此，反射包括输入、输出和中枢三个环节。有机体具有输入端（即感受器）、中枢和输出端（即效应器）三种结构，它们分别进行三种过程，即外界刺激的接收过程、通过中枢的传导过程以及输出端的输出过程，由这三个过程构成反射活动。

反射活动是神经整合的单元反应，每次反射是一个整合反应，缺少反射的神经活动不是完整的整合过程。在一个简单的反射（如膝跳反射）中就有整合作用。他引入协调的概念来讨论反射，协调在反射过程中起重要的作用，反射是对各种输入协调的结果。他还指出，中枢神经系统有兴奋过程和抑制过程，中枢的应答是整合性的。

他详细讨论了脊髓反射的各种特性,包括简单反射中的协调作用、反射之间的相互作用、复合反射中的同时性结合、复合反射中的继时性结合、反射的适应反应等。

在研究反射时,他发现了交互神经支配的神经协调方式。当一块肌肉收缩时,另一块与之相关的肌肉就放松。从神经活动来说,当支配肌肉收缩的运动神经元兴奋时,支配另一与之相关肌肉放松的神经元就被抑制。神经的兴奋性活动伴随着抑制性活动,这就是交互神经支配。

在突触水平上,他讨论了突触的整合作用,并首次提出了突触的概念。突触是一个神经元的末梢和另一个神经元的树突或胞体的接触点,它们是神经系统信号传递的基础。突触将许多输入转变为一个输出,这就是突触的整合作用。他指出,在突触部位有兴奋和抑制的相互作用,每一个突触都是一个协调机构。

在神经细胞水平上,他讨论了单个神经细胞的整合作用。一个运动神经元的整合作用表现为细胞对信号的整合,一个运动神经元能够整合兴奋输入和抑制输入,对各种信号进行评估,从而决定行为。因此他把单个神经元看作是整合的细胞基础,可以从单个运动神经元来看整体的脑的整合作用。

在动物整体水平上,他指出多细胞动物的个体不仅仅是许多器官的集合,动物体内的神经系统通过整合作用使分散的各个器官统一为具有一致性的动物个体。

他不但在中枢反射的层次上讨论了神经协调活动,而且把这些原理应用到较高层次的过程。例如,在感觉过程方面,他讨论过双眼视觉现象。

他研究了大脑皮质的整合作用，指出脑是一个完整的整体。他还讨论过心智与身体的整合，认为神经系统最高水平的整合是心智与身体的整合。

在研究方法方面，他对不同的研究方法进行整合，把神经解剖学、神经生理学以及行为研究等各种方法结合起来，进行神经系统的研究。

一般集成论的由来

一般集成论的观念是我在科研和教学第一线的长期实践中，在思想中逐步形成的。

20世纪80年代，我在老一辈科学家的鼓励下，开始关心物理学和生物学与医学的交叉研究，以后又对物理学和脑科学与认知科学的交叉研究产生了兴趣。在此期间，我结识了许多生物学领域、医学领域和神经科学领域的朋友，和他们进行过广泛的讨论，并在实验和理论方面得到过他们的热情帮助。

在20世纪90年代，由于国家科研工作的需要，我和合作者先后进行了国家攀登计划"核医学和放射治疗中先进技术的基础研究"项目的研究，以及国家自然科学基金重大项目"发展近场技术、研究生物大分子体系特征"的研究。

核医学和放射治疗是把核技术应用于医学临床的诊断和治疗，让核技术为人民健康服务。"核医学和放射治疗中先进技术的基础研究"项目包括核医学、放射性药物和放射治疗三个研究方向。这个项目的研究工作涉及核物理学、放射化学、医学、药理学、计算机科学等学科，有许多不同学科的研究工作者参与了这项研究。

"发展近场技术、研究生物大分子体系特征"项目是研究和发展纳米探测和操纵技术，把它们应用于生物大分子体系的研究，以了解生物大分子体系的特征。这个项目的研究工作涉及物理学、分子生物学、化学、纳米技术、精密机械技术等学科，也有许多不同学科的研究工作者参与了研究。

我在参与这两个研究项目工作的过程中，学习到了许多医学物理学和生物物理学的知识和技术。同时，通过生物学、医学、物理学等多学科交叉研究，我开始形成知识集成和技术集成的观念，并且体会到学科交叉研究中集成的重要性。

学科交叉并不是几个学科中原有课题和设备的拼凑，也并不是几个学科的研究人员表面上的组合。实质性的学科交叉需要不同学科的研究工作者为解决同一个科学技术问题打成一片，进行长期的合作研究。多学科交叉的过程是集成过程，需有不同学科的知识集成和技术集成，还要有合作团队的资源集成和管理集成。

当时还因为要把医学影像技术应用于脑功能成像研究，我开始学习脑科学和认知科学。人类的脑是自然界中最复杂的物质。脑科学的研究包括探测脑、认识脑、保护脑、开发脑、仿造脑等领域，每一个领域的研究又涉及许多学科。脑功能成像是用影像技术无损伤地测量人在静息状态和认知过程中脑区的活动和连接情况，从而了解脑的工作原理。进行脑功能成像的实验，若只有单个学科的知识或单种实验技术，则是不够的，需要有多学科的知识和实验技术的集成。

在学习脑科学的过程中，谢林顿的《神经系统的整合作用》一书给我很大影响。他说，中枢神经系统的主要作用是整合作用，动物中枢神经系统的整合作用使分散的各种器官统一为具有一致性的动物个体。谢林顿通过脊髓反射过程对中枢神经系统的整合作用进行了深入的研究。他还讨论了突触的整合作用和单个神经细胞的整合作用。

他的思想使我尝试用集成（即整合）作用的观点考察脑内不同层次的各种活动。事实表明，从分子、基因、突触、神经细胞、神经回路、功能专一性脑区、功能子系统到整体的脑，脑的不同层次都存在各种不同的集成现象。

在脑的系统水平上，脑是四个功能系统，即维持觉醒的功能系统、加工信息的功能系统、调节控制的功能系统和评估-情绪的功能系统的集成，这些系统的协调运作保证了脑的正常活动。除脑的结构与功能集成外，脑内还存在信息集成和心理集成等许多类型和多种形式的集成作用和集成过程。

当时，在进行脑功能成像实验之外，我和合作者还开展了神经信息学的工作。神经信息学是神经科学和信息科学相结合的交叉学科。神经信息学的一个方面是以信息和信息处理的观点，研究神经系统信息的载体形式，神经信息的产生、传输和加工，以及神经信息的编码、存储和提取的机制等，也就是研究脑的信息集成。

神经信息学的另一个方面是利用现代化的信息工具，将脑的不同层次的研究数据集中起来，建立神经信息数据库和神经信息工作平台，对数据进行分析、处理和建模，进行科学数据的交换、共享和合作研究。许多学科领域的科学数据共享，包括我们推动的神经科学数据共享，都是科学信息的集成和网络通信的集成。

2001年初，我全职到浙江大学工作，组织了学校的脑和智能研究中心。那时由于学校学科建设的需要，我和同事们先后组织和进行了"十五""211工程"重点学科建设项目"脑与认知科学及其应用"的研究，以及"985"二期项目"语言与认知研究"的工作。

"脑与认知科学及其应用"项目对脑与认知科学中若干问题和应用开展多学科的研究，其中包括分子神经生物学、神经-内分泌-免疫网络、神经信息学、认知科学应用等方面的研究。"语言与认知研究"项目对语言与认知的若干前沿问题进行多学科的研究，其中包括心智与意识、语言认知、社会认知、认知科学应用等方面的研究。

那时浙江大学还成立了语言与认知研究中心，这是学校哲学社会科学的创新基地。参加这个中心的工作，使我有机会向人文社会科学的专家们学习，和他们进行广泛的讨论，共同促进文理交融，也使我获得了进行自然科学和人文社会科学集成研究的一些经验。

在上面两个项目的研究中，我做过一些具体的研究课题。我从现有的实验事实出发，先后探讨过心智的本质问题、心理学的理论体系问题以及认知科学理论的集成问题。在心理学和认知科学领域中的这些探讨，促进了我对不同领域中各种集成现象的思考。

心智是非常复杂的现象。心智活动包括觉醒成分、认知成分、情感成分、意志成分等许多成分。我体会到，要对这些成分和它们之间的相互作用进行集成研究，还要对心智、脑、身体、自然环境、社会环境等进行集成研究，才能全面地了解心智的本质。

当代心理学包括许多基础学科和应用学科，它们分别有许多分支学科。当代心理学的基础学科有实验心理学、生物心理学、生理心理学、神经心理学、心理物理学、认知心理学、发展心理学、人格心理学、社会心理学等，当代心理学的应用学科有教育心理学、学校心理学、医学心理学、临床心理学、工程心理学、工业及组织心理学、体育心理学、军事心理学、广告心理学、司法心理学等。我提出了基于心理相互作用及其统一理论的心理学大统一理论，尝试把上述大量的不同学科集成到一个统一的理论体系之中。

当代认知科学有许多种研究取向，主要是认知的神经生物学研究取向、信息加工研究取向、具身认知的研究取向、情境认知的研究取向、社会认知的研究取向、进化心理学研究取向、发展心理学研究取向、人工智能研究取向等。我认为，认知科学的研究要面对现有的所有研究取向，通过对它们进行集成，来构建认知科学的集成理论。

近十多年来，我在几个单位和同事们合作，并且指导研究生，从事静息态和认知的脑功能成像的实验，以及医学影像技术的发展和应用，特别是同北京大学医学物理和工程北京市重点实验室进行了长期的合作。在这个实验室进行的工作，在硬件方面包括CT、MRI、SPECT、PET、fMRI、脑电等多种技术的发展和应用，在软件方面包括图像分割、配准、融合等多种方法的发展和应用，同时还进行放射治疗学的研究。

我们进行过结构成像和功能成像的研究，后来又推动分子影像学的研究。分子影像学是医学影像学和分子生物学相互交叉产生的学科，分子影像技术是利用各种医学影像技术对人体内部特定的分子进行无损伤的实时成像。

这些工作加深了我对集成现象的认识。我体会到，技术集成能够促进新技术的发展。在医学影像技术领域中，这种技术集成表现为多种模态影像技术、多种影像实验数据以及多种影像分析方法等的综合应用。

为了教育学的学科建设，我在浙江大学还探索过智能问题，并且和浙江大学及浙江师范大学的教师们一起，进行了教育神经科学的讨论和研究，因而逐步接触了智能集成和教育集成等观念。

智能包括心智能力、行为能力以及各种具体能力，智能具有层次性的结构和动态发展的过程。我认为，智能是心智能力和行为能力的集成，也是各种心理相互作用能力的集成；需要研究层次性的智能结构中的各种集成作用，以及动态的智能活动中的各种集成过程。

教育神经科学是神经科学和教育科学相结合的交叉学科。在教育神经科学领域中，一方面要进行面向教育理论和教育实践的神经科学研究，另一方面要发展基于脑科学研究成果的教育理论和教育实践。教育神经科学需要神经科学和教育科学等方面知识的集成。

在上面提到的科研和教学第一线的实践中，我接触到了许多不同领域的各种集成现象，因而在思想中逐步形成了一般集成论的观念。我注意到，在自然界、工程技术领域和人类社会中广泛存在多种多样的集成现象。这些现象的普遍性使我进一步考察各种集成作用和集成过程的特点以及它们的一般特性。

在向脑学习和研究不同领域相关实验事实的基础上，我归纳各种集成作用和集成过程的一些一般性的概念，如优化、全局化、互补、协调、符合、同步、绑定、涌现、适应、同化、集大成、大统一等概念，用来描述集成现象的共性。

我认为，有必要在探讨各种集成现象一般特性的基础上，总结各个不同领域中有关集成现象的事实和观念，构建一门研究各类集成现象一般特性和规律及其应用的学科。我把这门学科定名为一般集成论，它的英文名称命名为 general integratics，简称 integratics。

我还考虑把一般集成论应用于各种具体领域，构建一系列研究各种具体领域中集成现象特别是集成作用和集成过程的具体规律及其应用的子学科。我把这些子学科称为专门集成论，它们的英文名称命名为 special integratics。

这些专门集成论形成一个学科群。一般集成论是这个学科群中各种子学科的集成。这些专门集成论的种类很多，下面举出其中几种作为例子：

在生物领域中有各种集成现象，需要建立一门研究生物领域中集成现象的特性和规律及其应用的子学科，可以把它定名为生物集成论，它的英文名称是 bio-integratics。

在医学领域中有各种集成现象，需要建立一门研究医学领域中集成现象的特性和规律及其应用的子学科，可以把它定名为医学集成论，它的英文名称是 med-integratics。

在心理领域中有各种集成现象，需要建立一门研究心理领域中集成现象的特性和规律及其应用的子学科，可以把它定名为心理集成论，它的英文名称是 psycho-integratics。

在认知科学领域中有各种集成现象，需要建立一门研究认知领域中集成现象的特性和规律及其应用的子学科，可以把它定名为认知集成论，它的英文名称是 cogno-integratics。

在信息科学领域中有各种集成现象，需要建立一门研究信息领域中集成现象的特性和规律及其应用的子学科，可以把它定名为信息集成论，它的英文名称是 info-integratics。

在地球科学领域中有各种集成现象，需要建立一门研究地球科学领域中集成现象的特性和规律及其应用的子学科，可以把它定名为地球集成论，它的英文名称是 geo-integratics。

在空间科学领域中有各种集成现象，需要建立一门研究空间科学领域中集成现象的特性和规律及其应用的子学科，可以把它定名为空间集成论，它的英文名称是 space integratics。

在环境科学领域中有各种集成现象，需要建立一门研究环境科学领域中集成现象的特性和规律及其应用的子学科，可以把它定名为环境集成论，它的英文名称是 environment integratics。

在工程领域中有各种集成现象，需要建立一门研究工程领域中集成现象的特性和规律及其应用的子学科，可以把它定名为工程集成论，它的英文名称是 engineering integratics。

在技术领域中有各种集成现象，需要建立一门研究技术领域中集成现象的特性和规律及其应用的子学科，可以把它定名为技术集成论，它的英文名称是 technology integratics。

在教育领域中有各种集成现象，需要建立一门研究教育领域中集成现象的特性和规律及其应用的子学科，可以把它定名为教育集成论，它的英文名称是 education integratics。

在经济领域中有各种集成现象，需要建立一门研究经济领域中集成现象的特性和规律及其应用的子学科，可以把它定名为经济集成论，它的英文名称是 economics integratics。

在文化领域中有各种集成现象，需要建立一门研究文化领域中集成现象的特性和规律及其应用的子学科，可以把它定名为文化集成论，它的英文名称是 culture integratics。

在社会领域中有各种集成现象，需要建立一门研究社会领域中集成现象的特性和规律及其应用的子学科，可以把它定名为社会集成论，它的英文名称是 social integratics。

同时还需要建立许多具体领域中的专门集成论。例如，神经集成论（neuro-integratics）、脑集成论（brain integratics）、知识集成论（knowledge integratics）、智能集成论（intelligence integratics）、管理集成论（management integratics）等。

开展一般集成论和各种专门集成论的研究，需要许多学科的专家共同的、长期的努力。因此，我先写一本阐述一般集成论理论及其应用的专著，向各方面专家请教，希望通过相互切磋，逐步形成比较完整的理论体系。这就是本书的由来。

如前所述，一般集成论的观念是受谢林顿《神经系统的整合作用》一书的思想影响而逐步形成的。从分析脑内不同层次的集成现象出发，再进而逐步考察自然界、技术领域和人类社会中的集成现象；从提出脑集成论、神经集成论和仿脑学理论，逐步发展到提出一般集成论。

在此期间，我阅读了前人控制论、系统论、信息论以及综合集成法等方面的论著，这些理论对我很有启发。我分析了一般集成论和这些理论之间的联系，也弄清了一般集成论和这些理论之间的区别。希望我们开展的一般集成论以及一系列专门集成论的研究，能够补充和丰富前人这些理论的成果。

一般集成论的原理和方法

集成现象与一般集成论

在自然界、人的思维和实践以及人类社会中，存在着各种各样的集成现象。

什么是集成？复杂事物内部包含许多成分，通过它们之间的各种相互作用以及它们和环境之间的各种相互作用，集成为协调活动的统一体。不同层次的复杂事物有不同种类的内部集成成分、集成作用和集成环境，形成不同内容和不同形式的集成统一体。集成是动态的过程，集成过程中在一定条件下涌现新的特性。

在人们的日常生活中随时随地都有集成现象。例如，人在说话时，把一些相关的词组成一句话，这是集成现象；人在一定场合中，把所见到的各种事物组成统一的场景，这是集成现象；人在发信件时，把有关的事情和想法写成一封信，这是集成现象。在上述这些集成过程中既有有意识的信息加工，又有大量无意识的信息加工，是有意识活动和无意识活动的集成。

复杂事物存在多种集成作用和集成过程。自然界包括无生命的物理世界、能量、有生命的生物世界和人类思维的精神世界。在无生命的物理世界中，存在物质、能量、结构、功能、信息、运动等多方面的集成现象。在有生命的生物世界中，存在生物体物质、能量、结构、功能、信息、生命活动、生物进化等多方面的集成现象。在人的心智活动中，有感知、记忆、思维、智能、情绪、意识等多方面的集成现象。在人类社会的经济、政治、文化、教育等领域中都存在集成现象。在经济和政治领域中，有经济活动的集成现象和政治活动的集成现象。在文化领域中，有文化集成现象。在教育领域中，有多元教育的集成现象。在科学、技术、工程领域中，有知识、学科、技术、工程等多方面的集成现象。

集成现象是普遍存在的。我们要研究各种集成现象的共同特性和一般规律。一般集成论是研究自然界、人的思维和实践以及人类社会中各种集成现象共同特性与一般规律及其应用的一门科学。

各个专门的领域中存在不同的集成成分、集成作用、集成环境、集成过程和集成统一体等。各种专门集成论是将一般集成论应用于各个具体领域，研究具体领域中的集成现象的特性和规律及其应用的学科。

人们对各种自组织现象已经有过许多研究。自组织是自发的集成过程。与自组织理论相比，一般集成论的特点是：一般集成论不仅考察自发的集成，而且更加强调主动的、有控制的集成，对集成过程进行主动的控制、干预、优化和组织。从这个意义上说，一般集成论更加注重人工的集成。

一般集成论的原理

我们从各种集成现象的特性,归纳得到一般集成论的三个规律:一般集成论的多元与统一规律、一般集成论的层次与涌现规律、一般集成论的发展与优化规律。

(1) 一般集成论的多元与统一规律

复杂事物存在复杂性和多样性。无数不同的事物之间不同的相互作用以及它们的种种运动变化,构成了丰富多彩的世界。复杂事物是多元的、统一的集成体。它们既有多元结构,又有内部统一性。

复杂事物的多元特性指它们是多元的集成。它们内部并不是单纯的一种成分,而是有复杂的多种成分,各种成分有各自的特性。复杂事物并不是内部各种成分的简单叠加,而是通过相互作用而集成的统一整体。复杂事物的统一特性指它们内部多种成分共同存在并相互作用。多种成分是互补的,它们的协调运作形成复杂事物总体的活动。

以人的心智现象中的意识活动为例。意识是十分复杂的现象,是人脑内部多种因素、多个方面的集成。从神经生物学看,意识的神经基础是许多神经系统集成的神经网络。从心理学看,意识是许多心理要素集成的心理活动,又是心智、身体、环境的集成。从信息学看,意识是大量信息集成的复杂统一体,意识活动是有意识的信息加工和无意识的信息加工的集成。从进化论看,意识是脑在长期进化过程中集成优化的功能。从社会文化学看，意识是人的经验、文化、社会的集成。

(2) 一般集成论的层次与涌现规律

复杂事物存在层级现象，复杂事物并不是单一层次，而是有多个层次。对于有多个层次的集成体，较复杂的上面层次是由较简单的下面层次集成的。复杂事物的不同层次各自有不同的结构和运动，有各自的特点和规律。在多个不同层次之间存在相互作用。

在下面层次的成分集成为上面层次的集成体的过程中，涌现会发生在达到临界条件之时。涌现是指集成过程中突现出原来各个成分并不具备的新特性。临界条件是集成过程中涌现新特性所需要的条件。

以生命现象的集成层次为例。在生命现象中存在生物分子、基因、亚细胞结构、细胞、组织、器官、生物个体、生物群体、生态系统等许多不同的层次。生命现象的每一个层次都各有不同的集成现象，出现不同的涌现。例如，生物分子集成为活细胞的过程中涌现活细胞的各种生命活动，包括活细胞内部的物质输运、能量代谢、信息传递等，它们是单个生物分子所不具备的新的特性。

(3) 一般集成论的发展与优化规律

一切事物都处于不断的动态的发展之中。集成体是在集成过程中形成和发展的。

事物在集成过程中不断优化。集成体内部存在比较、竞争、选择、重建，会使事物不断完善。因此，事物的集成过程也是优化的过程。优化的集成是有效的集成。

以生命现象的生物进化为例。生物体在遗传基因的基础上与环境长期相互作用而进化。进化像是"修补匠"，生物在进化中不断优化。现今的生物体是生物在长期进化过程中经过自然选择的结果。

一般集成论的方法

一般集成论为我们观察复杂事物提供了新的观点和处理复杂事物的新方法。基于一般集成论的多元与统一规律，我们可以用多元与统一的观点观察事物，用合理还原和有机整合的方法解决问题。基于一般集成论的层次与涌现规律，可以用层次与涌现的观点观察事物，用层次分析和涌现观察的方法解决问题。基于一般集成论的发展与优化规律，可以用发展与优化的观点观察事物，用动态分析和优化处理的方法解决问题。

(1) 合理还原和有机整合的方法

在研究复杂事物时，合理还原和有机整合这两种方法都不可缺少。即把复杂的集成体还原到它的各种成分，考察事物的全部成分而不仅仅是其中一个成分，并在此基础上加以整合。我们既要研究事物内部的各个单元，又要研究大量单元构成的复杂网络。

对复杂的事物进行还原的研究时，我们要对集成体的各种成分分别进行分析研究。这种还原不是无限制的还原，而是还原到适当的、合理的、有意义的层次，因此称为合理还原。

对还原的成分进行整合的研究时，我们要把各种成分集成起来。这种整合不是简单叠加，而是了解事物内部各种成分之间的相互作用，以及它们和环境之间的相互作用，找出集成体内部各种成分之间的有机联系，按照事物本身的特性进行整合，因此称为有机整合。

以细胞研究为例。我们可以把细胞还原为细胞核系统、细胞质系统、细胞膜系统和细胞骨架系统。细胞核系统是细胞的遗传物质系统。细胞质系统含有生物分子、离子、水分子等，它们集成为多种细胞器及整个细胞。细胞膜系统包括质膜、核膜等，细胞内还有许多具膜的结构。细胞骨架系统的主要成分是微丝、微管、中间纤维等。在活细胞内，这四个系统有紧密的联系。活细胞的整体是由这些系统集成的复杂系统，这些系统的协调运行，保证了活细胞正常的生命活动。

(2) 层次分析和涌现考察的方法

对复杂事物要考察集成的层次现象和集成过程中的涌现现象。我们既要研究不同集成层次的不同组织和运动，又要研究集成过程中在一定临界条件下涌现的新特性。

对复杂事物进行层次分析时，我们要了解复杂事物存在哪些层次；不仅要考察一个层次，还要考察各个层次；分别研究各个层次的特点和规律，以及不同层次之间的关系。

对复杂事物进行涌现观察时，我们要了解不同层次出现的各种不同的涌现现象。特别要研究事物在集成过程中，在何种临界条件下涌现哪些新的特性，以及这些新的特性是如何涌现的。

以意识研究为例。脑和心智非常复杂，具有多个层次和多个维度。要从脑和心智的不同层次的各种成分的特性和相互作用，从它们与身体、环境和社会的相互作用，从它们的集成过程和涌现现象等方面，来对意识进行研究。从心理学的角度看，意识是意识觉醒、意识内容、意识指向、意识情感等要素的集成。从神经生物学的角度看，意识是觉醒程度、信息加工、全局广播、注意增强等要素的集成。从信息学的角度看，意识是身体与自我信息、环境信息、社会信息、文化信息等成分的集成。意识的集成理论是意识的心理集成、神经集成、信息集成与"心-身-环境"集成之集大成。

(3) 动态分析和优化处理的方法

集成是动态的过程。集成体是在集成体内部各种成分之间的集成作用下，以及集成体和环境之间的集成作用下发展的。要对复杂事物不断变化的集成过程进行考察，对集成作用下的发展进行动态的分析。

要了解事物的不同发展阶段的特点，我们在处理问题时就要分阶段、有步骤地进行集成。前面已经提到，一般集成论不仅研究自发的集成现象，而且强调主动的集成过程。

集成的过程是优化的过程。对于复杂事物，要在全面了解情况的基础上提出处理问题的多种方案，对它们反复进行评估和比较，选择最优方案，并在实施中不断调整。在集成过程中要对事物主动地、有计划地进行优化。

以知识集成为例。知识需要不断积累，对一个复杂的科学问题的研究，常分为若干阶段，循序渐进。知识集成是一个探索的过程，在长期实践中不断地增加知识、总结经验、改正错误，才能得到新的成果。在知识集成过程中，不但要把现有的知识组织起来，而且要通过集成来产生新的概念和问题，提出新的假设和计划，再通过新的实验检验来发展原有的知识。这就是知识创新的过程。

参考文献

中文文献

Lambright, W.H. (2009). 重大科学计划实施的关键: 管理与协调(王小宁 译). 北京: 科学出版社.

L. 贝塔朗菲. (1987). 一般系统论: 基础·发展·应用(秋同, 袁嘉新 译). 北京: 社会科学文献出版社.

N. 玻尔. (1964). 原子论和自然的描述(郁稻 译). 北京: 商务印书馆.

P.W. 齐纳, R.L. 约翰逊. (1986). 集合论初步(麦卓文, 麦绍文 译). 北京: 科学出版社.

包含飞. (2003). 生物医学知识整合论(一). 医学信息: 医学与计算机应用, 16(6), 274-279.

常杰, 葛滢. (2005). 统合生物学纲要. 北京: 高等教育出版社.

陈捷娜, 吴秋明. (2007). 产业集群的集成论阐释. 科技进步与对策, 24(3), 58-61.

陈宜张. (2008). 神经科学的历史发展和思考. 上海: 上海科学技术出版社.

戴汝为. (2009). 基于综合集成法的工程创新. 工程研究: 跨学科视野中的工程, 1(1), 46-50.

方嘉琳. (1982). 集合论. 长春: 吉林人民出版社.

费尔迪南·德·索绪尔. (1980). 普通语言学教程(高名凯 译). 北京: 商务印书馆.

弗朗西斯·克里克. (1994). 狂热的追求: 科学发现之我见(吕向东, 唐孝威 译). 合肥: 中国科学技术大学出版社.

顾凡及, 梁培基. (2007). 神经信息处理. 北京: 北京工业大学出版社.

海峰, 李必强, 冯艳飞. (2001). 集成论的基本范畴. 中国软科学, 1, 114-117.

海峰, 李必强. (1999). 管理集成论. 中国软科学, 3, 86-87.

赫伯特·金迪斯, 萨缪·鲍尔斯等. (2005). 走向统一的社会科学: 来自桑塔费学派的看法(浙江大学跨学科社会科学研究中心 译). 上海: 上海人民出版社.

胡启勇. (2002). 文化整合论. 贵州民族学院学报: 哲学社会科学版, 1, 36-40, 53.

黄秉宪. (2000). 脑的高级功能与神经网络. 北京: 科学出版社.

李必强, 胡浩. (2004). 企业产权集成论. 理论月刊, 5, 165-166.

刘晓强. (1997). 集成论初探. 中国软科学, 10, 103-106.

吕叔湘. (1979). 汉语语法分析问题. 北京: 商务印书馆.

牛世盛. (1997). 生命整合论: 生命现象探索. 北京: 中央民族大学出版社.

潘菽. (1985). 人类的智能: 人类心理图说. 上海: 上海科学技术出版社.

彭聃龄. (2001). 普通心理学. 北京: 北京师范大学出版社.

皮亚杰. (2006). 结构主义(倪连生, 王琳 译). 北京: 商务印书馆.

钱学森. (1986). 关于思维科学. 上海: 上海人民出版社.

钱学森. (2007). 创建系统学. 上海: 上海交通大学出版社.

钱学森, 于景元, 戴汝为. (1990). 一个科学新领域: 开放的复杂巨系统及其方法论. 自然杂志, 1, 4.

宋晓兰, 陈飞燕, 唐孝威. (2007). 无意识活动与静息态脑能量消耗. 应用心理学, 13(1), 33-36, 43.

宋晓兰, 唐孝威. (2008). 意识全局工作空间的扩展理论. 自然科学进展, 18(6), 622-627.

唐孝威. (1985). 正负电子对撞实验. 北京: 人民教育出版社.

唐孝威. (1992). 关于有丝分裂后期染色体作用力的讨论. 自然科学进展, 2, 454-456.

唐孝威. (1993). L3合作实验. 自然科学进展, 3, 303-308.

唐孝威. (1999). 脑功能成像. 合肥: 中国科学技术大学出版社.

唐孝威. (2001). 核医学和放射治疗技术. 北京: 北京医科大学出版社.

唐孝威. (2003a). 脑功能原理. 杭州: 浙江大学出版社.

唐孝威. (2003b). 意识的四个要素理论. 应用心理学, 9(3), 10-13.

唐孝威. (2004). 意识论: 意识问题的自然科学研究. 北京: 高等教育出版社.

唐孝威. (2005). 关于心理学统一理论的探讨. 应用心理学, 11(3), 282-283.

唐孝威. (2007). 统一框架下的心理学与认知理论. 上海: 上海人民出版社.

唐孝威. (2008a). 心智的无意识活动. 杭州: 浙江大学出版社.

唐孝威. (2008b). AMPLE智力模型: PASS智力模型的扩展. 应用心理学, 14(1), 66-69.

唐孝威. (2010). 智能论: 心智能力和行为能力的集成. 杭州: 浙江大学出版社.

唐孝威, 杜继曾, 陈学群, 魏尔清, 徐琴美, 秦莉娟. (2006). 脑科学导论. 杭州: 浙江大学出版社.

唐孝威, 郭爱克. (2000). 选择性注意的统一模型. 生物物理学报, 16(1), 187-188.

唐孝威, 黄秉宪. (2003). 脑的四个功能系统学说. 应用心理学, 9(2), 3-5.

唐孝威, 刘国琴. (1992). 花粉管生长和内部颗粒运动的定量测量. 植物学报, 34, 893-898.

唐孝威, 沈小雷, 何宏建. (2008). 关于人体经络的一个试探性观点. 中国工程科学, 10(11), 14-17.

唐孝威, 吴义根, 单保慈, 曾海宁. (2001). 神经元簇的层次性联合编码假设. 生物物理学报, 17(4), 806-808.

维纳. (2007). 控制论: 或关于在动物和机器中控制和通信的科学(郝季仁 译). 北京: 北京大学出版社.

韦钰, P. Rowell. (2005). 探究式科学教育教学指导. 北京: 教育科学出版社.

沃尔特·J.弗利曼. (2004). 神经动力学: 对介观脑动力学的探索(顾凡及, 梁培基等 译). 杭州: 浙江大学出版社.

阎隆飞, 唐孝威, 刘国琴. (1994). 花粉管胞质颗粒运输的流动轨道系统. 自然科学进展, 4, 599-602.

杨宁, 陈忠才, 卢萍, 张传茂, 翟中和, 唐孝威. (2003). 非洲爪蟾卵非细胞体系中星体的组装及其在核膜重建中的作用. 科学通报, 48(15), 1623-1627.

喻红阳, 李海婴, 吕鑫. (2005). 网络组织集成论. 理论月刊, 2, 110-112.

约翰·C.埃克尔斯. (2004). 脑的进化: 自我意识的创生(潘泓 译). 上海: 上海科技教育出版社.

张香桐. (1997). 张香桐科学论文集: 1936—1997. 上海: 中国科学院上海分院图书馆.

外文文献

Adeva, B., Aguilar-Benitez, M., Akbari, H., Alcaraz, J., Aloisio, A., Alvarez-Taviel, J., ... Zoll, J. (1990). The construction of the L3 experiment. Nuclear Instruments and Methods in Physics Research Section A: Accelerators, Spectrometers, Detectors and Associated Equipment, 289(1-2), 35-102.

Alkire, M.T., Haier, R.J., Barker, S.I., & Shah, N.K. (1995). Cerebral metabolism during propofol anesthesia in humans studied with positron emission tomography. Anesthesiology, 82(2), 393-403.

Alkire, M.T., Haier, R.J., Shah, N.K., & Anderson, C.T. (1997). Positron emission tomography study of regional cerebral metabolism in humans during isoflurane anesthesia. Anesthesiology, 86(3), 549-557.

Alkire, M.T., Pomfrett, C., Haier, R.J., Gianzero, M.V., Chan, C.M., Jacobsen, B.P., & Fallon, J.H. (1999). Functional brain imaging during anesthesia in humans: Effects of halothane on global and regional cerebral glucose metabolism. Anesthesiology, 90(3), 701-709.

Anderson, J.R., Dan, B., Byrne, M.D., Douglass, S.A., Lebiere, C., & Qin, Y. (2004). An integrated theory of the mind. Psychological Review, 111(4), 1036-1060.

Anderson, J.R. (1983). The architecture of cognition. Massachusetts: Harvard University Press.

Arnold, M.B. (1960). Emotion and personality. New York: Columbia University Press.

Baars, B.J., & Franklin, S. (2003). How conscious experience and working memory interact. Trends in Cognitive Sciences, 7(4), 166-172.

Baars, B.J. (1983). Conscious contents provide the nervous system with coherent, global information. In Davidson, R.J., Schwartz, G.E., & Shapiro, D. Consciousness and self-regulation. New York: Plenum Press.

Baars, B.J. (1988). A cognitive theory of consciousness. New York: Cambridge

University Press.

Barlow, H.B. (1972). Single units and sensation: A neuron doctrine for perceptual psychology? Perception, 1(4), 371-394.

Beggs, M.J., & Plenz, D. (2004). Neuronal avalanches are diverse and precise activity patterns that are stable for many hours in cortical slice cultures. The Journal of Neu-roscience, 24(22), 5216-5229.

Biswal, B.B., Mennes, M., Zuo, X.N., Gohel, S., Kelly, C., Smith, S.M., ... Milham, M.P. (2010). Toward discovery science of human brain function. Proceedings of the National Academy of Sciences, 107(10), 4734-4739.

Brooks, R.A. (1991). Intelligence without reason. In Mylopoulos, J., Reiter, R. Proceedings of the 12th International Joint Conference on Artificial Intelligence. San Mateo, CA: Morgan Kaufmann, 569-595.

Brooks, R.A. (1999). Cambrian intelligence: The early history of the new AI. Massachusetts: The MIT Press.

Buzsáki, G. (2006). Rhythms of the brain. New York: Oxford University Press.

Buzsáki, G. (2007). The structure of consciousness. Nature, 446(7133), 267.

Cacioppo, J.T., Berntson, G.G., Adolphs, R., Carter, C.S., Davidson, R.J., McClintock M.K., ... Taylor, S.E. (2002). Foundations in social neuroscience. Massachusetts: The MIT Press.

Chomsky, N. (1957). Syntactic structure. The Hague: Mouton.

Corbetta, M., Miezin, F.M., Dobmeyer, S., Shulman, G.L., & Petersen, S.E. (1991). Selective and divided attention during visual discriminations of shape, color and speed: Functional anatomy by positron emission tomography. Journal of Neuroscience, 11, 2383-2402.

Crick, F., & Koch, C. (2003). A framework for consciousness. Nature Neuroscience, 6, 119-126.

Crick, F. (1984). Function of the thalamic reticular complex: The searchlight hypothesis. Proceedings of the National Academy of Sciences, 81(14), 4586-4590.

Das, J.P., Naglieri, J.A., & Kirby, J.R. (1994). Assessment of cognitive processes: The PASS theory of intelligence. Boston: Allyn and Bacon.

Dehaene, S. (2001). The cognitive neuroscience of consciousness. Massachusetts: The MIT Press.

Denmark, F.L., & Krauss, H.H. (2005). Unification through diversity. In Sternberg, R.J. Unity in psychology: Possibility or pipedream? Washington, DC: American

Psychological Association.

Desimone, R., & Duncan, J. (1995). Neural mechanisms of selective visual attention. Annual Review of Neuroscience, 18: 193-222.

Duncan, J., Ward, R., & Shapiro, K. (1994). Direct measurement of attentional dwell time in human vision. Nature, 369 (6478), 313-315.

Eckhorn, R., Bauer, R., Jordan, W., Brosch, M., & Reitboeck, H.J. (1988). Coherent oscillations: Amechanism for feature linking in the visual cortex. Biological Cybernetics, 60 (2), 121-130.

Edelman, G.M., & Tononi, G. (2000). A universe of consciousness: How matter becomes imagination. New York: Basic Books.

Flourens, P. (1960). Investigations of the properties and the functions of the various parts which compose the cerebral mass. In von Bonin, G. Some papers on the cerebral cortex. Springfield, IL.: Charles C. Thomas, 3-21.

Matthei, E.H., & Fodor, J. (1984). The modularity of mind: An essay on faculty psychology. Cambridge, MA: MIT Press.

Fox, M.D., Snyder, A.Z., Vincent, J.L., Corbetta, M., van Essen, D.C., & Raichle, M.E. (2005). The haman brain is intrinsically organized into dynamic, anticor-related functional networks. Proceedings of the National Academy of Sciences, 102(27), 9673-9678.

Frackowiak, R., Friston, K.J., Frith, C.D., Dolan, R.J., & Mazziotta J.C. (1997). Human brain function. San Diego: Academic Press.

Fransson, P. (2005). Spontaneous low-frequency BOLD signal fluctuations: An fMRI investigation of the resting-state default mode of brain function hypothesis. Human Brain Mapping, 26(1), 15-29.

Fujii, H., Ito H., Aihara, K., Ichinose, N., & Tsukada, M. (1996). Dynamical cell assembly hypothesis: Theoretical possibility of spatio-temporal coding in the cortex. Neural Network, 9(8), 1303-1350.

Fuster, J. (1997). Network memory. Trends in Neurosciences, 20(1), 451-459.

Gallagher, H., & Frith, C. (2003). Functional imaging of 'theory of mind'. Trends in Cognitive Sciences, 7(2), 77-83.

Gardner, H. (1985). The mind's new science: A history of the cognitive revolution. New York: Basic Books.

Gardner, H. (1993). Multiple intelligence: The theory in practice, A Reader. New York: Basic Books.

Gazzaniga, M.S., Ivry, R.B., & Mangun, G.R. (1998). Cognitive neuroscience: The biology of the life. New York: W. W. Norton & Company.

Gazzaniga, M.S. (2000). The cognitive neuroscience (2nd ed.). Cambridge: The MIT Press.

Gibson, J.J. (1979). An ecological approach to visual perception. Boston: Houghton Mifflin.

Glassman, W.E. (2000). Approaches to psychology (3rd ed.). Philadelphia: Open University Press.

Goleman, D. (1995). Emotional Intelligence. New York: Bantam Books.

Gray, C.M., Knig, P., Engel, A.K., & Singer, W. (1989). Oscillatory responses in cat visual cortex exhibit inter-columnar synchronization which reflects global stimulus properties. Nature, 338, 334-337.

Greicius, M.D., Ben, K., Reiss, A.L., & Vinod, M. (2003). Functional connectivity in the resting brain: A network analysis of the default mode hypothesis. Proceedings of the National Academy of Sciences, 100(1), 253-258.

Greicius, M.D., & Menon, V. (2004). Default-mode activity during a passive sensory task: Uncoupled from deactivation but impacting activation. Journal of Cognitive Neuroscience, 16(9), 1484-1492.

Gusnard, D.A., & Raichle, M.E. (2001). Searching for a baseline: Functional imaging and the resting human brain. Nature Reviews Neuroscience, 2, 685-694.

Haberlandt, K. (1997). Cognitive Psychology (2nd ed.). Boston: Allyn and Bacon.

Hawkins, J.,& Blakeslee, S. (2004). On intelligence. New York: Henry Holt.

He, Y., Wang, J., Wang, L., Chen, Z.J., Yan, C., Yang, H., ... Zang, Y. (2009). Uncovering intrinsic modular organization of spontaneous brain activity in humans. PLoS ONE, 4(4), e5226.

Hebb, D. (1949). The organization of behavior: A neuropsychological theory. New York: John Wiley.

Hilgard, E.R. (1987). Psychology in America: A historical survey. New York: Harcourt Brace College Publishers.

Hillyard, S.A., & Picton, T.W. (1987). Electrophysiology of cognition. In Mountcastle, V.B., Plum, F., & Geiger, S.R. Handbook of Physiology: Vol 5. Baltimore: American Physiological Society.

Hothersall, D. (1984). History of psychology. Philadelphia: Temple University Press.

Jacob, F. (1977). Evolution and tinkering. Science, 196(4295), 1161-1166.

Kandel, E.R., Schwartz, J.H., & Jessell, T.M. (2000). Principles of neural science (4th ed.). New York: McGraw-Hill Medical.

Kosslyn, S.M., & Koenig, O. (1995). Wet mind: The new cognitive neuroscience. New York: The Free Press.

Kuhn, T. (1970). The structure of scientific revolution (2nd ed.). London: Cambridge University Press.

L3 Collaboration. (1993). Results from the L3 experiment at LEP. Physics Reports, 236(1-2), 1-146.

Lakoff, G., & Johnson, M. (1999). Philosophy in the flesh: The embodied mind and its challenge to Western thought. New York: Basic Books.

Lashley, K.S. (1929). Brain mechanisms and intelligence. Chicago: University of Chicago Press.

Laureys, S. (2005). The neural correlate of (un) awareness: Lessons from the vegetative state. Trends in Cognitive Science, 9(12), 556-559.

Lazarus, R.S. (1993). From psychological stress to the emotion: A history of changing outlooks. Annual Review of Psychology, 44, 1-22.

Le Doux, J. (1996). The emotional brain: The mysterious underpinning of emotional life. New York: Simon and Schuster.

Liljenström, H., & Århem, P. (2008). Consciousness transitions: Phylogenetic, ontogenetic, and physiological aspects. Amsterdam: Elsevier.

Logothetis, N.K., Pauls, J., Augath, M., Trinath, T., & Oeltermann, A. (2001). Neurophysiological investigation of the basis of the FMRI signal. Nature, 412(6843), 150-157.

Luria, A.R. (1966). Human brain and psychological processes. New York: Harper and Row.

Luria, A.R. (1973). The working brain: An introduction to neuropsychology. New York: Basic Books.

Maddock, R.J. (1999). The retrosplenial cortex and emotion: New insights from functional neuroimaging of the human brain. Trends in Neurosciences, 22(7), 310-316.

Maier, S.F., & Watkins, L.R. (1998). Cytokines for psychologists: Implications of bidirectional immune-to-brain communication for understanding behavior, mood, and cognition. Psychological Review, 105(1), 83-107.

McCann, S.M., Lipton, J.M., Sternberg, E.M., Chrousos, G.P., Gold, P.W., & Smith,

C.C. (1998). Neuroimmunomodulation: Molecular aspects, integrative systems and clinical advances. Annals of the New York Academy of Sciences, vol.840.

MeCarthy, R.A., & Warrington, E.K. (1990). Cognitive neuropsychology: A clinical introduction. London: Academic Press.

Mckiernan, K.A., Kaufman, J.N., Kucera-Thompson, J., & Binder, J.R. (2003). A parametric manipulation of factors affecting task-induced deactivation in functional neuroimaging. Jourmal of Cognitive Neuroscience, 15(3), 394-408.

Melmed, S. (2001). Series Introduction: The immuno-neuroendocrine interface. Journal of Clinical Investigation, 108(11), 1563-1566.

Miall, R.C., & Robertson, E.M. (2006). Functional imaging: Is the resting brain resting? Current Biology, 116(23), 998-1000.

Miller, G. (2007). Hunting for meaning after midnight. Science, 315 (5817), 1360-1363.

Naccache, L. (2006). Is she conscious? Science, 313(5792), 1395-1396.

Naglieri, J.A., & Das, J.P. (1990). Planning, attention, simultaneous and successive (PASS) cognitive processes as a model for intelligence. Journal of Psychoeducational Assessment, 8(3), 303-337.

Newell, A., & Simon, H. (1972). Human problem solving. Englewood Cliffs, NJ: Prentice-Hall.

Newell, A. (1990). Unified theories of cognition. Massachusetts: Harvard University Press.

Northoff, G., Bermpohl F. (2004). Cortical midline structures and the self. Trends in Cognitive Sciences, 8(3), 102-107.

Piaget, J. (1983). Piaget's theory. In Mussem, P. Handbook of child psychology, vol.1(4th ed.). New York: Wiley.

Posner, M.I., & Rothbart, M.K. (2006). Educating the human brain. Washington DC: American Psychological Association.

Purves, D., Augustine, G.J., Fitzpatrick, D., Katz, L.C., LaMantia A.S., & McNamara, J.O. (1997). Neuroscience. Sunderland: Sinauer Associates.

Raichle, M.E., MacLeod, A.M., Snyder, A.Z., Powers, W.J., Gusnard, D.A., & Shulman, G.L. (2001). A default mode of brain function. Proceedings of the National Academy of Sciences, 98(2), 676-682.

Raichle, M.E., & Mintun, M.A. (2006). Brain work and brain imaging. Annual Review of Neuroscience, 29, 449-476.

Raichle, M.E., & Snyder, A.Z. (2007). A default mode of brain function: A brief history of an evolving idea. Neuroimage, 37(4), 1083-1090.

Raichle, M.E. (2006). The brain's dark energy. Science, 314(5803), 1249-1250.

Salovey, P., & Mayer, J.D. (1990). Emotional intelligence. Imagination, Cognition and Personality, 9(3), 185-211.

Schiff, N.D., Ribary, U., Moreno, D.R., Beattie, B., Kronberg, E., Blasberg, R., ... Plum, F. (2002). Residual cerebral activity and behavioural fragments can remain in the persistently vegetative brain. Brain, 125(6), 1210-1234.

Sdorow, L. (1995). Psychology (3rd ed.). Madison: Brown and Benchmark.

Searle, J. (2000). Consciousness. Annual Review of Neuroscience, 23, 557-578.

Shannon, C.E., & Weaver, W. (1949). The mathematical theory of communication. Urbana: Univesity of Illinois Press.

Sherrington, C.S. (1906). The integrative action of the nervous system. New York: Charles Scribner's Sons.

Shulman, G.L., Corbetta, M., Buckner, R.L., Fiez, J.A., Miezin, F.M., Raichle, M.E., ... Petersen, S.E. (1997). Common blood flow changes across visual tasks:. Decreases in cerebral cortex. Journal of Cognitive Neuroscience, 9(5), 648-663.

Shulman, R.G., Hyder, F., & Rothman, D.L. (2004). Energetic basis of brain activity: Implications for neuroimaging. Trends in Neuroscience, 27(8), 489-495.

Singer, W., & Gray, C. (1995). Visual feature integration and the temporal correlation hypothesis. Annual Review of Neuroscience, 18, 555-586.

Staats, A.W. (1999). Unifying psychology requires new infrastructure, theory, method, and research agenda. Review of General Psychology, 3(1), 3-13.

Sternberg, R.J., & Grigorenko, E.L. (2001). Unified psychology. American Psychologist, 56(12), 1069-1079.

Sternberg, R.J. (1985). Beyond IQ: A triarchic theory of human intelligence. New York: Cambridge University Press.

Sternberg, R.J. (2005). Unity in psychology: Possibility or pipedream? Washington, DC: American Psychological Association.

Tang, X. (1998). Meshwork supported fluid film model of cell membranes. Chinese Physics Letter, 15(10), 770-771.

Thompson, E., & Varela, F.J. (2001). Radical embodiment: Neural dynamics and consciousness. Trends in Cognitive Science, 5(10), 418-425.

Tian, L., Jiang, T., Liu, Y., Yu, C., Wang, K., Zhou, Y., ... Li, K. (2007). The relationship within and between the extrinsic and intrinsic systems indicated by resting state correlational patterns of sensory cortices. Neuroimage, 36(3), 684-690.

Tononi, G., Sporns, O., & Edelman, G.M. (1994). A measure for brain complexity: Relating functional segregation and integration in the neurons system. Proceedings of the National Academy of Sciences, 91(11), 5033-5037.

Treisman, A., & Gelade, G. (1980). A feature-integration theory of attention. Cognitive Psychology, 12(1), 97-136.

Treisman, A., Sykes, M., & Gelade, G. (1977). Selective attention stimulus integration. In Dornie, S. Attention and performance VI. Hilldale NJ: Lawrence Erlbaum, 333-361.

Underleider, L., & Mishkin, M. (1982). Two cortical visual systems. In Ingle, D., Mansfield, R.J.W., & Goodale, M.S. Analysis of visual behavior. Cambridge MA: MIT Press, 549-586.

Varela, F.J., Thompson, E., Rosch, E., & Kabat-Zinn, J. (1991). The embodied mind: Cognitive science and human experience. Massachusetts: The MIT Press.

Vogeley, K., May, M., Ritzl, A., Falkai, P., Zilles, K., & Fink, G.R. (2004). Neural correlates of first-person perspective as one constituent of human self-consciousness. Journal of Cognitive Neuroscience, 16(5), 817-827.

Von Bertalanffy, L. (1950). An outline of general system theory. British Journal for the Philosophy of Science, 1(2): 139-165.

Von Bertalanffy, L. (1976). General system theory: Foundation, development, applications (Rev. ed.). New York: George Braziller.

Von der Malsburg, C. (1981). The correlation theory of brain function. Goettingen: Max-Planck-Institute for Biophysical Chemistry.

Waelti, P., Dickinson, A., & Schultz, W. (2001). Dopamine responses comply with basic assumptions of formal learning theory. Nature, 412, 43-48.

Wickelgren, I. (2003). Tapping the mind. Science, 299 (5606), 496-499.

Wilson, E. (1998). Consilience: The unity of knowledge. New York: Alfred A. Knopf.

Waelti, P., Dickinson, A., & Schultz, W. (2002). Brain-computer interfaces for communication and control. Clinical Neurophysiology, 113(6), 767-791.

Zeki, S. (2003). The disunity of consciousness. Trends in Cognitive Sciences, 7(5), 214-218.

Zeki, S. (1993). A vision of the brain. Oxford: Wiley-Blackwell.

名词简释

大统一心理学：将心理学各个分支领域统一起来的理论

地球集成论：研究地球科学领域中集成现象的特性和规律及其应用的学科

仿脑学：仿造脑的学科

工程集成论：研究工程领域中集成现象的特性和规律及其应用的学科

合理还原：还原到合适的层次

环境集成论：研究环境科学领域中集成现象的特性和规律及其应用的学科

集成论：研究集成现象的特性和规律及其应用的学科

技术集成论：研究技术领域中集成现象的特性和规律及其应用的学科

教育集成论：研究教育领域中集成现象的特性和规律及其应用的学科

经济集成论：研究经济领域中集成现象的特性和规律及其应用

的学科

空间集成论：研究空间科学领域中集成现象的特性和规律及其应用的学科

脑集成论：研究脑内集成现象的特性和规律及其应用的学科

脑与环境集成：脑和环境的集成过程

认知集成论：研究认知领域中集成现象的特性和规律及其应用的学科

认知整合理论：用整合观点研究认知的理论

社会集成论：研究社会领域中集成现象的特性和规律及其应用的学科

神经集成论：研究神经系统集成现象的特性和规律及其应用的学科

生物集成论：研究生物领域中集成现象的特性和规律及其应用的学科

文化集成论：研究文化领域中集成现象的特性和规律及其应用的学科

细胞集成论：研究细胞的集成现象的特性和规律及其应用的学科

心理集成：心理过程的集成

心理集成论：研究心理领域中集成现象的特性和规律及其应用的学科

心理相互作用：心理现象中的相互作用

信息集成论：研究信息科学中集成现象的特性和规律及其应用

的学科

一般集成论：研究集成现象一般特性和规律及其应用的学科

艺术集成论：研究艺术领域中集成现象的特性和规律及其应用的学科

医学集成论：研究医学领域中集成现象的特性和规律及其应用的学科

有机整合：根据要素的联系进行整合

智能集成：心智能力、行为能力及其他各种具体能力的集成

智能集成论：研究智能领域中集成现象的特性和规律及其应用的学科

知识集成论：研究知识领域中集成现象的特性和规律及其应用的学科

专门集成论：研究各个具体领域中集成现象的特性和规律及其应用的学科

General Integration Theory

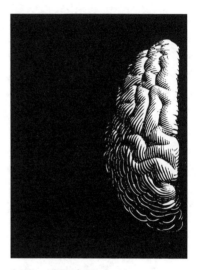

Author: TANG Xiaowei
Translator: WANG Dahui
Proofreader: WANG Xiaolu

ZHEJIANG EDUCATION PUBLISHING HOUSE · HANGZHOU

Preface

Our brain is the most complex system in nature and its activity is the most complex form in all kinds of material movement. One important subject in contemporary science and technology is to study, understand and apply the mechanism of the brain. This book will propose the general integration theory based on our knowledge about the brain.

While investigating the structure and function of the brain, we are exploring it at different levels, including molecule, gene, synapse, neuron, neural circuit, brain region with specific function, the functional system of the brain, the integrated system of the brain, mind and behavior. At the microscopic and mesoscopic levels, there exist different kinds of integrative unity in our nervous system, including synapse, neuron and neural circuit, etc.; while at the macroscopic level, there are brain functional systems and the integrated brain, etc.

In the brain, there are different integrative components at different levels. Various ways of interactions contribute to integrative unities with various forms and functions. At different levels of the brain, there are many types and forms of integrative action and integration process. These integrative unities at different levels of the brain further integrate into unified, integrated brain with complex structures and functions, showing colorful mind and generating various behaviors. The integration inside the brain is a dynamic and evolving process, accomplished by integrative actions inside the brain or between the brain and the environment.

The brain provides us with a laboratory of researching integrative actions and integration processes. Based on the experimental facts on the brain, this book discusses the properties of types and forms of integrative actions and processes inside the brain from the perspectives of structural and functional integrations, informational integration, and psychological integration. It suggests that we need a new discipline to exclusively investigate the properties, principles and applications of integration phenomena in the brain, which can be named as brain integratics. This discipline primarily conducts researches at the system level of our brain.

After the integrative actions and integration processes of the nervous system are investigated, another discipline that studies the properties, principles and applications of integration phenomena in nervous system needs to be established, which can be called neuro-integratics. This discipline involves research on integration phenomena at all levels of the nervous system, especially at microscopic and mesoscopic levels.

If we shift our attention away from the brain, we notice the following facts: integrative actions and integration processes not only are of vital importance to brain activities, but also exist extensively in nature, science, technology and human society. In all these fields at different levels, integration components of different types interact with each other and integrate into integrative unities of different levels and types, and generate new properties under some conditions. Integrative actions and integration processes in these fields have special rules respectively, but they also have some common characteristics, which can be described with some general concepts.

Based on the knowledge about the brain and the related experimental observations in various fields, this book summarizes some general concepts about integrative actions and integration processes, such as optimization, globalization, complementarity, coordination, conformation, synchronization, binding, emergence, adaptability, assimilation, global integration, grand unification, etc., all of which can be used to describe the common characteristics of various integration phenomena.

Although people have known the concept of integration for a long time and have separately discussed a few concrete integration phenomena, they have not conducted a unified and systematic research on the integration phenomena in nature, science, technology and human society, including integrative actions and integration processes. Based on the fact that there exist all kinds of integration phenomena in the aforementioned fields, this book proposes to establish a new discipline, conducting researches on integration phenomena of different fields, including the general principles and applications of integrative actions and integration processes. This discipline is termed in this book as general integration theory or general integratics.

This book consists of three parts. The first part is concerned with integration phenomena and integration theories of the brain. In this part, we also introduce some experimental facts about the brain, investigate the integrative actions and the integration processes of brain, and finally propose to establish the disciplines of neuro-integratics, brain integratics and brainics.

The second part discusses the general integration theory. Derived from the fact that various integrative actions and integration processes exist extensively in different fields, we try to summarize the general concepts of integration phenomena, to propose the arguments of general integration theory, and to explain its connections with system theory and the differences between them.

The third part discusses the applications of general integration theory. It has been applied in various fields, such as biology and medical science, psychology and behavioral science, knowledge innovation and discipline construction, engineering and technology, culture and education, etc.

As the author of this book is just a beginner in neuroscience and cognitive science, and knows only a little about engineering technology, humanities and social sciences, there might be some mistakes and flaws in the book. I wish the experts in these fields could tolerate the ignorance of a beginner and correct them without hesitation. I also invite readers to give their suggestions and criticisms about this book.

This book is funded and supported by Department of Science and Technology of Zhejiang Province.

Contents

Part I Integration Phenomena in the Brain and Integration Theory of the Brain 001

Chapter 1 The Integrated Brain 003
1.1 Structure and Function of the Brain 004
1.2 The Complex Network of and the Communication in the Brain 011
1.3 Brain Activity and Energy Consumption 017

Chapter 2 Integration of the Brain 023
2.1 Structural and Functional Integration of the Brain 024
2.2 Information Integration in the Brain 027
2.3 Psychological Integration in the Brain 034

Chapter 3 Neuro-integratics, Brain Integratics and Brainics 043
3.1 Neuro-integratics 043
3.2 Brain Integratics 045
3.3 Brainics 048

Part II General Integration Theory 051

Chapter 4 Integration Phenomena in Different Fields 053
4.1 Integration Phenomena in Nature 053

4.2 Integration Phenomena in Technology 056
4.3 Integration Phenomena in Cross-disciplinary Studies 059
4.4 Integration Phenomena in Human Society 061

Chapter 5 Exploring General Integration Theory 064
5.1 General Integration Theory 064
5.2 Globalization and Modularity 073
5.3 Reduction and Synthesis 074
5.4 Binding and Association 076
5.5 Reconstruction and Optimization 077
5.6 Criticality and Emergence 079
5.7 Complementarity and Coordination 081
5.8 Coincidence and Synchronization 082
5.9 Adaptability and Assimilation 085
5.10 Global Integration and Grand Unification 086

Chapter 6 Characteristics of General Integration Theory 088
6.1 General Integration Theory Based on Learning From the Brain 088
6.2 General Integration Theory and Special Integration Theory 091
6.3 Relation and Difference Between General Integration Theory, System Theory and Other Theories 095

Part III Applications of General Integration Theory 101

Chapter 7 Bio-integratics 103
7.1 Bio-integratics and Biological Evolution 104
7.2 Integration in a Living Cell 106
7.3 Integration in Human Body 110

Chapter 8 Psycho-integratics 118
8.1 Theory of Mental Interactions and Psycho-integration 119
8.2 Integration for Consciousness 125
8.3 Integration for Cognition 133

8.4 Mental Integration 136
8.5 Integration of Mind and Behavior 144

Chapter 9 Knowledge Integratics 147

9.1 Knowledge Integration and Cross-disciplinary Studies 148
9.2 Integrative Model of Selective Attention 155
9.3 Integration in Psychological Science 158
9.4 Integration of Theories in Cognitive Science 163

Chapter 10 Engineering Integratics 172

10.1 Integration in "Big Science" 173
10.2 Integration of Large Scale Experimental Facilities 175
10.3 Integration in Medical Imaging Techniques 179

Chapter 11 Education Integratics 185

11.1 Integrated Intelligence 185
11.2 Integration of Contents of Education 194

Appendix 197

References 219

Terminology 233

Epilogue 237

Part I

**Integration Phenomena in the Brain
and Integration Theory of the Brain**

The brain is the most complex system in nature and its activity takes the most complex form of material movement. Based on large amounts of experimental facts, the integration phenomena of the brain can be observed, and the integration theory of the brain may be proposed.

Starting from the experimental facts, this part illustrates the integrative entity of the brain, and further explores all sorts of integrative actions and integration processes existing in the brain. Built on these, discipline and brain integratics, is proposed to guide the research on integration phenomena and principles on the brain. Meanwhile, the disciplines of neuro-integratics and brainics are also established.

This part consists of three chapters. Chapter 1 discusses the integrated brain; Chapter 2 elaborates on the integration of the brain; and Chapter 3 proposes the establishment of disciplines of neuro-integratics, brain integratics and brainics.

Chapter 1　The Integrated Brain

We first explore the most complex system in nature: the human brain.

The human brain has hierarchical structures: from biomacromolecule, gene, subcellular structure, nervous cell (neurons and glia cells), assembly of neurons, neural circuit, brain region with specific functions, and functional system of brain to the whole brain. These structures of the brain are the basis of its activities. Thus, to study this complex system, we need to approach each level to unveil its structure, function and mechanism.

The complexity of the brain is not only manifested by the complexity of brain hierarchies, brain structures, brain functions, brain circuitry, brain communications, brain interactions with the environment, but also by the advanced and colorful functions of brain-mind phenomena and human behaviors.

This chapter discusses the integrated brain, and introduces some experimental facts about the brain from the perspectives of structure and function of the brain, complex network and communication in the brain, brain activity and energy consumption, which prove that the brain is an integrated unity.

1.1 Structure and Function of the Brain

The hierarchies of the brain roughly consist of three levels: microscopic level, mesoscopic level and macroscopic level. From small to large, at microscopic level, there are molecule, subcellular structure, and nervous cells; at mesoscopic level, there are assembly of neurons and neural microcircuitry; at macroscopic level, there are brain regions with specific functions, functional system of the brain and the whole brain. Microscopic level in the brain refers to the level of molecule and cell levels, far from that in the physics. Nervous structures at different levels in the brain have different spatial scales. At microscopic, mesoscopic and macroscopic levels, the spatial scales differ greatly. For example, the order of magnitude of nerve cell body is about 10^{-6} meters, while that of the brain region with specific functions is about 10^{-2} meters. The processes of activities of nervous structures at different levels have different time scales. At microscopic, mesoscopic and macroscopic levels, the time scales also differ greatly. For example, the electric impulse of neuron has an order of magnitude of about 10^{-3} seconds, while the process of the functional system of the brain has an order of magnitude of about 10^{-1} seconds.

Various experimental techniques can be used to study the structures and functions of different levels of the brain, for instance, the technique of molecular neurobiology, neurogenetics, neuroanatomy, electro-neurophysiology, microscopic imaging, brain structure imaging and brain function imaging and so on.

The technique of brain function imaging can be used to study the macro-functions of the brain, and can form dynamic images with no harm to human bodies. The methods we have adopted rely on the techniques of functional Magnetic Resonance Imaging (fMRI), Single

Photon Emission Computerized Tomography (SPECT), Positron Emission Computerized Tomography (PET), Electroencephalography (EEG), Event-related Potentials (ERP), and Magnetoencephalography (MEG), etc.

Neuroscientists have extensively investigated the structures and functions of our brain and constantly make progress. Much has been unveiled about the brain (Purves et al., 1997; Gazzaniga et al., 1998; Kandel et al., 2000; Chen, 2008). However, because of the intricacy of the brain, there are still many problems to be solved.

At the microscopic level, in human brain, there are approximately 10^{11} multivariant neurons and more glia cells, forming about 10^{14}—10^{15} synaptic connections between neurons. The neuron consists of cell nucleus, cytoplasm, cytomembrane and cytoskeleton. The basic structure of cytomembrane is a lipid bilayer embedded with protein, with many protein complexes on the membrane, such as receptors and ion channels. The ion channel is the pathway of ions through the membrane.

Inside the neuron, there are all kinds of neurotransmitters and neuromodulators. The former is a chemical substance with a lot of variants, such as acetylcholine and L-glutamic acid, and functions in the transmission of neural signals. The latter is used to modulate the biochemical reactions of neurons of many sorts, such as dopamine and noradrenaline.

The neurons exist with many different shapes. They are cells with many branch structures: a large amount of tiny branches called dendrite through which the neurons can receive the neural signals; and there is usually a slender fiber called axon, which is responsible for transmitting the neural signals. Moreover, dendrites are of vital importance in neural information processing. Zhang Xiangtong (1997) conducted research on the physiological functions of dendrites.

The synapse is the structure for neurons to interact with each other. The neuron interacts with other neurons through large amounts of synapses, with one neuron on average forming $10^3 - 10^4$ synaptic connections. The synapse is dynamic and plastic, playing a key role in neural information processing. Thus the neuronal network is very complex.

Inside the brain, the number of glia cells is about tenfold of that of neurons. They not only support and nurture the neurons, but also modulate the activities of neurons, affecting the function of neurons. In addition, the molecule emitted by the glia cell can regulate the contraction and relaxation of brain vessels, adjusting the local blood support for neural activities in the brain (Chen, 2008).

For neural structures and functions, the mesoscopic level is between microscopic level and macroscopic level while bridging these two levels. Some neurons and glia cells are connected into neural circuit at the mesoscopic level. The research on neural circuit is currently one of the hottest themes. Experimentally, there are many researches on the molecular and cellular mechanism of neural circuit with its functions and plasticity. Theoretically, there are many models of neural circuits (Freeman, 2004; Liljenström & Århem, 2008).

At the macroscopic level, the brain is inside the cranial cavity, protected by meninx. There is distribution of blood vessels with circulating blood, which provides nutrients for the brain's normal physiological activity and removes the waste. The complex brain consists of cerebrum, interbrain, midbrain, cerebellum, pons, and medulla; the combination of midbrain, pons and medulla is called the brainstem. The brain and the spinal cord are the central part of the nervous system, constituting the whole nervous system with the peripheral nervous system.

The brain has two hemispheres: left and right, connected by

fibers. The surface of the brain is bumpy with many rugae. The main gyruses divide the brain into four lobes: frontal lobe, parietal lobe, occipital lobe, temporal lobe and insula. The brain has grey matter and white matter, with the former consisting of nerve cells as cerebral cortex and the latter consisting of nerve fiber tracts under cerebral cortex. The oxygen and glucose for the brain all activities are provided by blood in the brain vessels.

The cerebral cortex includes many specific brain regions with various functions, such as visual cortex area, auditory cortex area, somatosensory cortex, motor cortex and Broca's area, etc. These regions constitute the functional system of the brain, such as the systems of receiving, manipulating and storing information; and these functional systems further make up the integrated brain.

In terms of the functional systems of the brain, we have proposed the theory of the four functional systems based on experiments. The essay *The Theory of Four Functional Systems of the Brain* (Tang & Huang, 2003) introduced Luria's theory (Luria, 1973) of three functional systems of the brain, and proposed the theory of four functional systems of the brain based on the Luria's theory. Some discussions in that essay are quoted as follows:

> Luria conducted clinical observation and rehabilitation training towards masses of brain-damaged patients. He observed that damaged parts of the brain might cause specific psychological disorder, that one function of the brain is not only related to one specific part of the brain but also related to other parts of the brain, for each part of the brain is strongly linked with other parts. Luria (1973) classified the brain into three intimately related functional systems based on experimental facts, and proposed the theory of three functional systems of the brain.

In the book *The Principle of Neuropsychology*, Luria exemplified that the brain has three functional systems as follows: the first one is to modulate and maintain the intensity and the state of consciousness, with related brain regions as brainstem reticular formation and limbic system; the second one is responsible for receiving, processing and storing information, corresponding brain regions are parietal lobe, occipital lobe, and temporal lobe; and the third one has the duty of formulating procedures, modulating and controlling mental activities and behaviors, with the frontal lobe as the related brain region. The whole human behavioral and mental activities are the result of the coordination of these three functional systems. This theory proposed by Luria is of vital importance to understand the overall functions of the brain.

However, we have noticed that besides these functions, evaluation and emotion, as necessary mental activities, should also have corresponding brain functions. Due to the limitation of experimental materials at that time, Luria failed to include the functional systems related with evaluation and emotion, while the current experimental facts are constantly proving the importance of these mental activities.

Seen from the experimental facts, the function of evaluation is common in many mental activities (Edelman & Tononi, 2000). During evolution the human organism has formulated a system which caters to individual and familial living and development by evaluating the input information from the outer environment. Based on the innate appraisal structure, the organism formulates the standards of evaluation combining the previous experience and the current needs; the evaluation system compares the significance of input information with the standard of appraisal, and then reports the result; the individual accepts or rejects or deals with the information and makes decisions on how to react to it on the basis of the degree of importance according to the reported result; the decisions in turn modulate the organism and react to the outer environment through the functional system which is responsible for modulating and controlling the mental activities and behaviors (Huang, 2000).

Inside the brain, each stage of information processing needs the appraisal on information, thus the evaluation inside the brain is ongoing mental activities. As the evaluation system has plasticity, the standards of evaluation are constantly forming and developing in the process of individual learning, and as a result they are constantly changing.

The evaluation-emotion functional system has similar organizational structures with the second and the third functional systems proposed by Luria. It is also a multilayered system. The emotion system is the fundamental part of the evaluation-emotion functional system, which evaluates the overall information in the context, and generates highlyintensive subjective experiences and reactions. In the higher order regions of the brain, the evaluation system can perform precise evaluation on special information, even concrete results of thinking.

The appraisal performed on the external information can also give rise to the individual emotional experience: the information which conforms to individual needs or wishes is positively evaluated and might generate positive emotional experience; while the information which does not conform to individual needs or wishes is negatively evaluated and might generate negative emotional experience (Arnold, 1960; Lazarus, 1993).

The amygdala inside the brain is important for memorizing the events related to praise and punishment. The dopaminergic neurons of ventral tegmental area and substantia nigra compact can react to the difference between expected praise and the actual praise (Waelti et al., 2001). This might also be a part of the evaluation system.

The limbic system, related to function of emotions (Le Doux, 1996), is also a part of evaluation-emotion functional system. Moreover, a certain part of the frontal lobe might be the higher order region for the evaluation-emotion functional system.

To make up for the deficiency of Luria's theory of three functional systems, which failed to include the evaluation-emotion functional system, we list the evaluation-emotion functional system as the fourth functional system. The reason is that evaluating the information and generating emotional experience are the basic functions of the brain; however, the previous three functional systems have not included the functions of evaluation and emotion; the evaluation-emotion functional system is different from the other functional systems, and thus it is necessary to include it as an independent functional system.

On the basis of previous discussion, we have developed Luria's theory of three functional systems and thus proposed the theory of four functional systems of the brain, acknowledging that there are four relatively independent and densely connected functional systems, including the first functional system — to modulate and maintain the intensity and the state of consciousness, the second functional system — to receive, process and store information, the third functional system — to modulate and control mental activities and behaviors, the fourth functional system— to evaluate information and generate emotional experience. These four functional systems are integrated into the whole brain, which interact and coordinate with each other, performing all kinds of behaviors and mental activities.

From the above explanation, it can be concluded that the complex human brain is an integrative system.

1.2 The Complex Network of and the Communication in the Brain

One important feature of the nervous system is the conduction and treatment of nerve signals. It has been deeply studied in the field of electro-neurophysiology at the microscopic level (Purves et al., 1997; Gu & Liang, 2007). The experiments proved that two neurons connected by synapses transmit nerve signals through the coupling of electric and chemical process.

The cell membrane of neurons has ion channels which are sensitive to voltage. There are different sorts of ions in the liquid inside and outside of neurons, such as sodium ions, potassium ions and chloride ions. When the neuron is in rest state, the concentration of ions inside the neuron membrane differs from that of ions on the outside, inducing a potential difference in and out of the neuron membrane. The potential inside the membrane is negative in contrast to the potential outside of the membrane, which is called resting potential.

When the neuron is stimulated and active, the sodium ions outside the membrane selectively pass through the membrane and enter the neuron, causing the change of membrane potential. In turn, it produces constantly changing potential called graded potential. When the membrane potential reaches a certain value, electrical impulses with a stereotype of waveform and amplitude will be generated and last for a short time, which is called action potential. This sort of action potential caused by changes of ionic transports of cellular membrane flows along the axon in one direction.

There is a cleft in the synapse between two neurons. When the action potential pulse flows to the presynaptic membrane of axon terminals, the transmitters in the vesicles in the presynaptic terminal will be released into synaptic cleft. Then they diffuse through the cleft, and arrive at the cellular membrane of another neuron, bonding with specific receptors on the membrane and causing the ion flow of the membrane of postsynaptic neurons.

There are many kinds of chemical synapses. The excitatory synapses make the postsynaptic membrane depolarization; the inhibitory synapses make the postsynaptic membrane hyperpolarization. If the potential of postsynaptic membrane depolarizes to a certain value, the action potential of postsynaptic neuron will be triggered.

The feature of action potential is "all or none", that is, either there is no action potential or the action potential is generated while its waveform and amplitude stay unchanged. In the process of information transfer, the impulse amplitude of action potential will not decline. The stimulus from outside can produce a series of action potentials, the pulse sequences of which carry the information of stimulus features.

At the macroscopic level, the communication inside the brain is proceeding in the network of the brain, integrated by functional subsystems. We have discussed it in the book *Principles of Brain Function* (Tang, 2003a). Several paragraphs are quoted from that book as follows:

> Neuroanatomy and neuropsychology have provided experimental evidence for functional subsystems of the brain. Sensors in each part of human body are projected to specific regions in sensory cortex of the brain while the specific regions in motor cortex controls the movement of each part of the body, thus the sensory system and motor system both have precisely located at the brain network to fulfill specific functions. On the aspect of higher-order cognitive

functions, the researches of Broca and Wernicke have exemplified that language has a series of separable and locatable subsystems (Kandel et al., 2000).

Frackowiak et al. (1997) pointed out that the separation and integration of the brain functions display the basic experimental fact. On the one hand, the brain consists of lots of functional subsystems, which processes information separately (Zeki, 1993). One kind of brain damage might influence one kind of brain function, but the others will not be affected (McCarthy & Warrington, 1990). On the other hand, the functional subsystems connect with each other to form an integrated neural network. The overall function of the brain relies on the interplay between the functional subsystems (Fuster, 1997). Large amounts of experiments have been carried out using brain imaging technology. These experiments further demonstrate that there are many separate but coordinating functional subsystems inside the brain.

These functional subsystems, which are relatively separate and mutually connected with different functions, are composed of large amounts of neural circuits inside the brain. One functional subsystem consists of related brain regions which help to perform the same function. There are circuits among the subsystems based on lots of neural pathways. In this way, many functional subsystems are connected with others in various ways and thus constituting the whole network of the brain (Huang, 2000). [Many scientists have conducted researches on the network modular organization of the spontaneous blood oxygen level dependent activities with the technology of functional magnetic resonance imaging (He et al., 2009).]

A functional subsystem can be equivalently described as a complex electronic circuit, which can transform, combine, analyze and record the input information. Here the input and output neural signals both refer to signals transferred between functional subsystems instead of physical stimulus from outer environment. Besides the route itself, this kind of equivalent electronic circuit includes the input terminal and output terminal of the signals, various kinds of gating structures for modulation, equivalent circuit connected with other functional subsystems, etc. A functional subsystem has input terminal and gated input, and can receive two types of inputs: one type of input signal passes through the input terminal and activates the brain regions of the functional subsystems; the other type of input signal passes through the gated input and modulates the brain activities of the functional subsystems, such as inhibition or enhancement.

A functional subsystem always has multiple input terminals. To be brief, we can just consider that only one input terminal has signal input, while the others do not. A functional subsystem also has multiple output terminals, of which we can also consider that only one is in use. The gating structure of functional subsystem mimics logic AND gate, NOT gate and OR gate in the electronic circuit, with the functions of identification, coincidence and anticoincidence. For example, when the gating function is inhibition, the input signal of the functional subsystem is inhibited; but when the gating function is coincident, the input signal of the functional subsystem is enhanced. All functional subsystems are widely connected by equivalent circuits, which are indispensable for the whole system.

In recent years, the default mode network of the brain has raised growing concern. We have introduced it in the essay *The Extended Global Workspace Theory of Consciousness* (Song & Tang, 2008). Here several paragraphs are cited as follows:

Recently, a generally proved phenomenon is that some brain

regions are more active when they are in the resting state or undergoing simple and passive sensory task of stimulations than when they are undergoing active stimulus-response tasks (Shulman et al., 1997; Raichle et al., 2001; Raichle & Snyder, 2007). When all kinds of tasks for processing outer information are performed, these brain regions usually manifest the deactivation state and the deactivation regions are consistent with more activated regions in the resting state. Moreover, whenever in the resting state or the working state, the spontaneous low-frequency blood oxygen level dependent signals in these regions are strongly varied with time, namely, they form a functional connected network (Raichle et al., 2001; Greicius et al., 2003; Fransson, 2005), which is called default mode network. Among them, the posterior cingulate cortex/ precuneus and the middle prefrontal cortex/the ventral anterior cingulate cortex are two core areas. In the resting state, the energy metabolism of these brain regions is the most active (Raichle & Mintun, 2006), demonstrating that some organized activities are undergoing in the resting state, although these activities are inhibited by external tasks and manifest the deactivation.

The pattern of activity inside the brain, which is caused by the default mode network organized by the way of functional connectivity is referred to as the default mode of activity of brain functions (Raichle & Snyder, 2007). The default mode of activity exemplifies that the brain is a highly self-organized organ which can be activated without stimulus from outside. The aforementioned default network is a special one. Based on the fact that the level of activity declines when there is a task, it is considered that in the resting state, the brain regions of default mode network are undergoing some kind of activities which are irrelevant with external stimulus. These activities are probably related to the processes of monitoring internal and external environment, forming self-consciousness, extracting episodic memory and maintaining the perception and mentality of the brain (Raichle & Snyder, 2007).

The default network is negatively correlated with some brain regions related to attention and working memory (Fox et al., 2005), and the load of cognition can modulate the intensity of activities of the default network, the level of activity decreases with the increasing of task difficulty (McKiernan et al., 2003). Some studies have also found that the spontaneous low-frequency blood oxygen level dependent signals of the default mode network are negatively correlated with the activities of sensory cortex widely spread in the posterior cortex, including visual, auditory, and somatic cortex (Tian et al., 2007). Although this is not consistent with the experiments by Fox et al., there might be an antagonism between the activities of default mode network and those of processing external information.

The brain regions of the default mode network participate widely in the brain activities. According to the result of activation compared with the control condition, we believe that these brain regions might be involved in the activities of episodic memory (Greicius & Menon, 2004), self-referential process (Northoff & Bermpohl, 2004), emotional process (Maddock, 1999), subjective feeling process and the first person perspective process (Vogeley et al., 2004; Gallagher & Frith, 2003) and conscious awareness maintenance (Raichle et al., 2001). Due to the complexity of cognitive processing in the resting state, no activation of a task can simultaneously "copy" these regions, thus the direct evidence of the functions of default mode network cannot be given. However, the stable deactivation network demonstrates the similarities of these complex cognitive processes. From the current evidence that these regions are far away from relatively elementary sensory and motor cortex, and that their activities are always restrained by external stimulus-response tasks, we can assume that they might be undergoing internal information processing irrelevant with external stimulus.

It can be seen from the above descriptions that on the aspects of brain complex network and communication in the brain, the complex brain is the integrated brain. To further understand the overall synchronization properties and the information transmission and integration of the brain, people are working on the avalanches in the neural network and the brain functional connectivity (Beggs & Plenz, 2004; Biswal et al., 2010).

1.3 Brain Activity and Energy Consumption

Brain activities include physiological activities and psychological activities. Inside the brain, there are such processes as blood circulation, oxygen metabolism, glucose metabolism and there are also nerve electric activities and neuronal chemical reaction. Besides all these physiological activities, the brain also has developed mental functions such as cognition, memory, learning and language, which belong to psychological activities. The physiological and psychological activities of the brain are intimately connected and consistent.

On the aspect of activities of brain regions, we have discussed the activation of brain regions and the interaction between brain regions (Tang, 2003a). One basic feature of a functional subsystem of the brain is the activation of relevant brain regions. A functional subsystem can be in different states. The resting state is the ground state while the active state (excited state) refers to the state when the brain is activated. The activation has different levels. The activation of the brain regions refers to the electrical activitiy of neurons in the regions and the related biochemical reactions. The parameter that describes the activation of brain regions is the activation level, measuring the overall activities in the brain regions, including the total effect of excitation, the level of biochemical reaction and the energy metabolism. In a functional subsystem, the activation of brain regions has a spatial distribution. At the system level of the brain, we only consider the overall level of activation of functional subsystems instead of its internal intricate activation distribution.

Another basic feature of the functional subsystem is its interaction with other subsystems. They transmit nervous signals mutually, including electric signals and biochemical signals of neurons. In the whole network of the brain, the complex interplay between functional subsystems are implemented by the signal transmission through the pathways between different subsystems. These signals are among the subsystems themselves instead of stimulus from external environment. The external physical stimuli are firstly transformed into nervous signals, and then delivered to the input terminals of the related functional subsystems. The parameter of the signals is the intensity of the signals, including the amplitude and duration of the signals. The input signal activates the brain regions of functional subsystems, then the brain regions will interact with neighboring regions by transmitting signals through the pathways. The parameter of describing the pathway is channel efficacy.

If we observe the functional subsystems of the brain and the whole brain network from the perspectives of activation and interaction, it seems that we could have a roughly unifying understanding of various brain functions such as sensation, perception, motion, attention, acquisition, memory, awareness and so on. For instance, the sensory organs are the input interface of the interplay between the sensory functional subsystem and the external environment, and the external physical signals are transformed into neural signals, which are transmitted into the sensory functional subsystem. The perception is the activation and interaction of sensory and perceptual functional subsystems by the top-down and bottom-up signals. The locomotive organs are the output interface between the motor functional subsystem in the brain and the external environment, and the signals from other subsystems are transmitted into the motor functional subsystems, acting on the movement organs, and generating and controlling motions. Attention is the interaction between controlling systems in the brain and controlled functional subsystems, enhancing the attended subsystems while suppressing the ignored subsystems. Learning is resulted from frequent activation of some subsystems, which makes the pathways of the network smoother. The store of memory is the formation of relevant structures in the subsystems, while the retrieval of it is the reactivation of these structures. Brain activities, conscious or not, are all brain activation and interaction, while the emergence of awareness corresponds to phase transition between relevant functional subsystems promoted by modulating and controlling systems.

From the perspective of information theory, the activities inside the brain are undergoing complex information processing (Huang, 2000). At the brain system level, the carrier of information inside the brain is the activated brain regions of functional subsystems. Inside the brain, regarding the information processing, the information receiving, coding, storage, retrieval, transmission, transformation and generation are all implemented by the activation of the brain regions and interaction between the brain regions.

We have discussed the energy consumption of the brain in the resting state in the essay *Unconsciousness and Energy Consumption of the Brain in the Resting State* (Song, Chen, & Tang, 2007). Some paragraphs are quoted here to exemplify it:

> Recently, the development of brain functional imaging techniques grants us the ability to study brain energy consumption under various conditions. The research demonstrates that the weight of the adult brain accounts for only 2% of the whole body, while it consumes 20% of the total energy, and is 10 times more than accounted by the proportion of weight. However, if the individual is doing a task, which needs stimulus from external environment, the increase of the energy consumption of the brain is no more than 1% (Raichle & Mintun, 2006), namely, the energy consumption of the brain in the resting state is far more than the extra energy consumption of doing a task.
>
> Thus, the intrinsic activities of the brain in the resting state not only manifest the low-frequency blood oxygen level dependent signals, but also consume large amounts of energy. The brain not only reacts passively to the external environment, but also undergoes self-organized activities. It has been pointed out that 80% of energy consumption in the brain is on neural signal processing (Shulman et al., 2004). We believe these intrinsic activities are essential information processing inside the brain.
>
> Previously, we discussed the conscious state when the individual is conscious and can perceive clearly. On the other hand, the experiments have shown that when the state of awareness changes, the energy consumption of the brain will change. For instance, the anesthesia caused by medicine can make the glucose metabolism of the whole brain decline on a large scale, from 31% to 68% with different medicines (Alkire et al., 1995; 1997; 1999). Another data are from vegetative patients who could not perceive and were

generally unconscious (Laureys, 2005). Their research found that the energy metabolism of the brain of vegetative patients is 30%-40% lower than the control group (Schiff et al., 2002). From the perspective of the brain reaction caused by external stimulus, most researches have shown that vegetative patients can only react to the stimulus by lower-level sensorimotor cortex. However, there are also case studies which found that some vegetative patients could react to language stimuli, with the experimental data of the fMRI showing that some could even manipulate mental images guided by the researchers (Naccache, 2006). Moreover, the materials about deep sleep show that at this stage, the activation level of the brain metabolism declines; meanwhile, PET and behavioral experiments show that at this stage, the brain is undergoing the activities which consolidate the memory and complement the missed information in the daytime (Miller, 2007).

Some possible explanations have been proposed about the energy consumption of the brain in the resting state (Raichle, 2006). For instance, the energy is used to support the spontaneous cognitive activities, like daydreams. However, these activities also require awareness and perception as ongoing tasks, which should consume the same amount of energy. Another explanation is that this part of energy is used to balance the neural connections of excitability and inhibition, which might be another cause of the brain's self-organized activities (Buzsaki, 2007). However, this kind of explanation has no solid evidence. The third explanation seems reasonable, that the intrinsic activities inside the brain are constantly maintaining the information to better deal with the change of environment. Although many explanations have been proposed, the nature of energy consumption of the brain in the resting state is still unclear. Therefore, people refer to it as the "dark energy" of the brain (Raichle, 2006).

We have once proposed an explanation for the mechanism of energy consumption (Song, Chen, & Tang, 2007). We believe that large amounts of energy consumption in the resting state are mainly used on unconscious processing without doing external tasks, besides the repair of neural cells and transmission of proteins which are suggested by Raichle and Mintum (2006). The energy consumption also includes the energy for preparation for environmental change. These processes will not stop with the emergence of external tasks, which occupy large amounts of attention resources.

We believe that although the content of unconscious activities cannot be perceived, the relevant brain regions are activated, but the activities are at a low level, which cannot be aware of. These information processing activities which make the brain regions activated need energy. Although a single brain region consumes little energy, lots of parallel unconscious activities consume lots of energy.

In these regards, without any external tasks, the brain does not rest (Miall & Robertson, 2006), instead, it keeps activated. Compared with the temporary activation by external tasks, the basic activities in the resting state are continuous (Gusnard & Raichle, 2001). The brain is undergoing continuous information processing, namely unconscious processing.

From the perspective of the brain's activities and energy consumption, the brain's function is implemented in the integration processes such as material integration, energy integration and information integration. The complex brain is an integrated brain.

Chapter 2 Integration of the Brain

The previous chapter introduces the basic knowledge of brain science, and based on the experimental facts of the brain and mentality it demonstrates that the brain is integrated. The integrated brain is closely connected with the brain integration. This chapter differs from Chapter 1 in which we exemplify the brain as an integrated unity with the view of integration while in this chapter we use the same view to illustrate various integration phenomena, especially various integrative actions and integration processes inside the brain.

The integration inside the brain is a dynamic process, during which different parts of the brain are integrated into a unified brain unity. This process is developing with time, realized by interactions inside the brain and interactions between the brain and the external environment. The various interactions of the brain not only integrate each part of the brain into a unified brain, but also manage the neural system by integrating each physiological system and organ of the body into a unified body.

Sherrington has conducted an innovative research on the integrative action of central nervous system. He discussed in detail about the reflex action of the spinal cord in his book *The Integrative Action of the*

Nervous System (Sherrington, 1906), explaining the integrative action of nervous systems at different levels. He stated that the function of the central nervous system is integrative action.

There are many kinds of integrative actions and integration processes in the brain. In the following sections, we will investigate the structural and functional integration of the brain, the information integration in the brain and the psychological integration in the brain.

2.1 Structural and Functional Integration of the Brain

This section discusses the integrative action and the integration process from the perspective of the structure and function of the brain. Structural integration refers to the integrative action and the integration process of brain structures, and functional integration refers to the integrative action and the integration process of brain functions. In biology, structure is the basis of function, thus the structural integration is closely correlated with the functional integration. The former provides the basis for the latter while the latter enhances the former.

These two processes of integration happen in different time scales with large spans. In terms of the evolution of species, the time scale of integrative actions spans from several hundred thousand years to several million years. In terms of an individual's whole lifespan, the time scale of integrative actions spans from several decades to more than one hundred years. In terms of his/her stage of learning process, the time scale spans from several days to several years. In terms of a certain type of mental activity, it spans from tens of milliseconds to several tens of seconds. In the following part we shall exemplify the structural integration and functional integration of the brain from the aspect of the evolving and ontogenetic process of the brain.

Throughout evolution and natural selection, especially by labor, human beings have not only developed their nervous system vital to survival, but also gradually formed complex functional systems of the brain. Eccles explained the evolution of the brain in the book *Evolution of the Brain* (Eccles, 2004). Lots of facts show that the structure and functions of the human brain have been gradually developing during the process of evolution. The process of evolution is the process of brain integratics. The present human brain is the result of long-term structural and functional integration.

The complex brain consists of a large number of neurons and glia cells, but how are they integrated into the complex brain? Purve et al. (1997) showed many materials about the development of the brain in the book *Neuroscience*, exemplifying the development from embryo to adult brain which involves many complex processes, such as the differentiation of nervous system, the segmentation of the brain, the neuronal migration, the formation of synapses and the construction of nervous circuits. These facts also demonstrate that in the process of brain development, there are all kinds of integration phenomena. The process of brain development is the integration process of the brain.

As mentioned above, the brain is a complex system with multiple levels, and there are many middle levels between biomacromolecules, nervous cells and the integrated brain. These middle levels are integrated by neurons and glia cells, and they in turn integrate into the brain.

The brain has plasticity, which manifests differently at different levels of the brain with different properties of time and space. In the following part we will discuss the plasticity of synapses, brain regions, the network inside the brain and the traumatic brain regions.

The synapse has plasticity. The experiment conducted by Hebb (1949) showed that when the synapse of neuron transmits excitatory signal and meanwhile the postsynaptic neuron transmits the signal, the efficacy of synaptic connection enhances. This enhancement in the process of nervous activity is called the plasticity of the synapse, which is the microscopic foundation of the plasticity of network inside the brain.

The brain region has plasticity. For any brain region, its activation leads the change of connection between this region and the other regions. For instance, some long-lasting training will make the reorganization of the regions relevant with the training and the expansion of related cortexes.

The complex network of the brain has plasticity, too. The plasticity of brain regions and their connections change the complex network of the brain. Some studies have proved that practical activities have a great influence on the development of network in the brain, such as training. The plasticity of the brain is the foundation of learning and memorizing.

In terms of the development of the brain, the axons and dendrites of neurons of a newborn brain increase fast, with the weight of the brain increasing constantly at the same time. So do the brains of a child and an adolescent. Posner and Rothbart (2006) studied the developing process of attention network inside children's brain.

The studies on the damage (lesion) of sensory organs indicate that if a certain kind of peripheral sensory organ is damaged, the corresponding regions of cortex will be reorganized, gradually losing their former functions and developing into sensory systems of other information. Such cases have proved that the cerebral cortex has plasticity.

The injured brain region also has plasticity. If a certain functional area in the cerebral cortex is partially damaged by traumatic brain injury, the injured brain region can be supplemented by other regions after recovery, namely, the system in one brain region that has the previous functions may be transferred to other regions.

To summarize, the brain is constantly reorganizing, which is also the process of structural integration and functional integration.

2.2 Information Integration in the Brain

The brain is an organ of dealing with information, thus communication inside the brain is an important feature of the brain. The information integration in the brain is an important aspect of integrative actions and integration processes. A large amount of information in the brain is integrated into the overall information. In this section we will discuss the integrative action and the integration process from the perspective of information processing.

The information integration in the brain is closely tied to structural integration and functional integration discussed in the previous section. The information integration is based on structural integration. In addition to the neural circuit constituted by neighboring neurons, there might be long range connections as well. These connections have adequate spatial distribution, laying the structural foundation for information integration in the brain.

The information integration is in turn the basis of functional integration. Sherrington (1906) pointed out that the integration of nervous systems is accomplished by the transmission of neural signals. The neural signals can transmit very fast and have adequate temporal distribution.

At the level of single neuron, one neuron can integrate the input excitatory signals and inhibitory signals. The illustration of integration of neural signals in Chapter 12 of *Principles of Neural Science* written by Kandel et al. (2000) is quoted as follows:

> Each neuron in the central nervous system, whether in the spinal cord or in the brain, is constantly bombarded by synaptic input from other neurons. A single motor neuron, for example, may be innervated by as many as 10,000 different presynaptic endings. Some are excitatory, others inhibitory; some strong, others weak. Some inputs contact the motor cell on the tips of its apical dendrites, others on proximal dendrites, some on the dendritic shaft, others on dendritic spines. The different inputs can reinforce or cancel one another.
>
> These competing inputs are integrated in the postsynaptic neuron by a process called neuronal integration. Neuronal integration reflects at the level of the cell the task that confronts the nervous system as a whole: decision making. A cell at any given moment has two options: to fire or not to fire an action potential. Charles Sherrington described the brain's ability to choose between competing alternatives as the integrative action of the nervous system (Sherrington, 1906). He regarded this decision-making as the brain's most fundamental operation.

The encoding of neural information, that is, the issue of how neural information is encoded, represented and processed in the brain is one of the basic problems of information integration. We have discussed the information encoding of neuronal clusters in the essay *Hypothesis of Hierarchically Associated Codingbased on the Neuronal Clusters* (Tang et al., 2001). Some of the contents are quoted as follows:

The research on information encoding of the brain has a long history. It began from the classical cell assembly hypothesis raised by Hebb in 1949 to Barlow's hypothesis of single neuron encoding in 1972, then to Fujii et al who proposed the hypothesis of dynamic temporal-spatial encoding of neural population in 1996. Different arguments are always competing with each other. Some of these arguments are listed here: Whether a single neuron or the neural assembly encode the stimulus information? Which carries the stimulus information, the emergence time of action potential of neurons or the average firing rate? Due to the high complexity of neural system, the experimental techniques available nowadays cannot be used to fully understand the principle of neural information encoding. The currently existing encoding theories have difficulties at different levels in explaining the working mechanism of neural system.

Many neurobiological experiments have suggested that neural system has several obvious features in information processing. Firstly comes diversity, namely to recognize different states or forms of the same target, such as different facial expressions of the same people at different times or ages, or some kind of objects having different colors, shapes and sizes. Secondly comes robustness, that is, the death and trauma of a single neuron do not cause lots of related neural information. Finally comes hierarchy, which means that there are different levels dealing with the stimulus information, including the level of characteristics and the level of abstraction. For instance, the details of your grandmother and the abstract concept of grandmother belong to two different levels.

Barlow's hypothesis (1972) of single neuron encoding has several basic points: the hypothesis of average firing rate encoding; the hypothesis of optimal stimulus; the hypothesis of each single neuron having a single function of continuity. The direct expansion of Barlow's theory on the aspects of higher order cognitive functions is the grandmother cell hypothesis. Obviously, the difficulty of this theory is combinatorial explosion, which derives from its fundamental hypothesis: there is a need for many cells of encoding which can represent the combinations of multiple features and properties of objective entities. However, the limitation of the number of brain cells cannot fulfill this need. As pointed out once by von der Malsburg in 1981, this theory generates more difficulties than the problems that could be solved. In addition, this theory cannot solve the issue of robustness, namely the problem of memory management brought by the generation and death of cognition cells.

The classical cell assembly hypothesis proposed by Hebb (1949) confronts two closely related problems: the problems of overlapping and binding. The problem of overlapping refers to the situation that when two stimuli arrive at the same time, the two related cell assemblies cannot be discerned because all neurons in the cell assembly increase their average distribution rate. The fundamental reason for this difficulty is that Hebb's classical cell assembly lacks of internal structures, while the external knowledge it should express is structural and hierarchical. The problem of binding is the problem related with the integration of information in multiple parallel pathways in the cortex. For example, when a red circle and a green triangle appear in the visual field meanwhile, the cell assemblies which represent red and green in the color region are activated at the same time and the cell assemblies which represent circles and triangles in the shape region increase their average firing rate at the same time, thus in the framework of Hebb's classical cell assembly, no accurate binding can be completed.

Apart from the aforementioned two encoding theories, there are also other influential theories, such as the hypothesis of primitive representation by Fujii et al. (1996), the hypothesis of encoding by synchrony oscillation (von der Malsburg, 1981), the place encoding theory, and the recently widely-concerned hypothesis of dynamic temporal and spatial encoding of neural assemblies, etc. It should be pointed out that some conclusive neurobiological experiments have proved that even in the same cortical region, there are completely different paradigms of encoding, thus we should consider the possibility that there should be two or more kinds of encoding in the same brain region.

To overcome the difficulties confronted by the existing encoding theories and to explain more experimental facts, we try to propose one hypothesis, which integrates the hypothesis of grandmother cell and Hebb's classical cell assembly hypothesis, and is called the hypothesis of hierarchical associated encoding of cluster of neurons. We believe that the basic unit of neural encoding is the cluster of neurons, which consists of a group of neurons with similar functions and positioning, that every cluster of neurons has selective expression of features and reacts most intensely towards the stimulus having these features while the reaction declines with the increase of differences between this stimulus and optimal ones, that the cluster of neurons which participate in the encoding have hierarchies, including the cluster encoding of abstract concepts and the cluster encoding of specific attributes, and that a basic feature of neuron cluster encoding is an associate expression, that is, clusters of different levels are bound together to jointly express complex stimulus.

This hypothesis takes into account not only the previously mentioned features of information processing, but also the following experimental facts: (1) there are functional columns in the cortex, and neurons within the same functional column have similar responses to specific stimulus; (2) an olfactory functional column reacts optimally to a specific odor molecule, and considerably response sensitively to similar molecules. The other functional columns have similar properties. However, the relations between the cluster of neurons which reacts to optimal stimulus and non-optimal stimulus need further research.

The concept of cluster of neurons is obviously different from that of cells of grandmother, for its basic unit is cluster instead of a single neuron. It is also different from Hebb's classical cell assembly, because the clusters of neurons have selective feature expressions and hierarchical associate encoding, while Hebb's classical cell assemblies are based on anatomical connection, organized with the principle of joint distribution. Although on some aspects the concept of clusters of neurons is similar with Hebb's classical cell assembly, the recognition of clusters of neurons is completed based on the joint activities of internal members inside a cluster, which is also the associate increase of discharge rate within a certain time window (for example: 100ms). All the anatomical positions and functions of neurons are similar inside a cluster of neurons.

The advantage of hierarchical associate encoding of clusters of neurons is to avoid the difficulty of the current encoding hypothesis. First of all, robustness can be achieved. Due to the fact that every cluster of neurons has many (10^3 or more) neurons, the trauma or death of a single neuron won't influence the expression of the whole cluster upon specific stimulus. Secondly, the hypothesis assumes that neuron clusters have internal structures and complex stimulus with many features, and attributes can activate a group of clusters which represent different features and attributes, and that their proper binding can jointly encode the complex stimulus. The joint expression needs specific binding, a possible mechanism is the selective attention of the more advanced brain region. As for the synchronous oscillation for binding, there needs further research. In addition, this kind of joint expression can work at different levels, that is, clusters which encode abstract concepts and concrete attributes can jointly express various complex stimuli. Finally, in this hypothesis, there are many clusters of neurons with similar functions which can solve the problem of diversity, for example, different facial expressions of grandmother can be encoded and jointly expressed by clusters of neurons.

At the level of the whole brain, the book *Principles of Brain Function* (Tang, 2003a) has discussed the integrative action of neural signals in the process of activation of brain regions and their mutual interactions: (1) the inactivated brain regions maintain the original state when they have not received input signals; the activation level of the activated brain regions declines with time when they do not receive input signals; (2) the input signals activate the brain regions and the activation level of the brain regions increases as the input signals become stronger and stronger; (3) the output signals from the activated brain regions reach the association areas of the brain and the signals become stronger and stronger as the activation levels of the activated brain regions and the efficiency of pathways increase; (4) the activated brain regions act on the joint brain regions, making the pathways between them more efficient, and the joint brain regions react on the activated brain regions.

Moreover, another aspect of integrative action and integration process must be mentioned, that is, the integration of brain, body and environment. The brain is not a closed system but an open system in the integrated unity of the body and the environment which includes natural environment and social environment.

The brain, body and environment cannot be separated, instead, the brain and body have to constantly interact with natural environment and social environment, which shape the brain and mind in these processes. In cognitive science, there are several different opinions about the interaction between the mind-brain system and environment, such as embodiment cognition (Lakoff & Johnson, 1999), context cognition (Brooks, 1999), dynamic system cognition (Thompson & Varela, 2001), etc. Various interactions among the brain, body and environment enhance the integration of the brain, body and environment.

2.3 Psychological Integration in the Brain

The mental activity is the advanced function of the brain. The psychological integration in the brain refers to the integrative action and integration process of mental activities. We have pointed out in the book *Unconscious Mental Activities* (Tang, 2008a) that mentality consists of arousal and attention, cognition, emotion, intention and their interactions. From that book we quote some relevant contents as follows:

> Mind has the component of arousal and attention. A certain level of arousal is the basis of mental activity. The individual cannot experience without arousal. There are different levels of arousal, which reflect the overall state of individual mentality. The level of arousal varies with time. Arousal also has something to do with the other components of mind, affected by emotion and related to intention.

Mind also has the component of cognition. The individual subjective experience has specific content. In the process of cognition, the individual knows what he perceives and the meaning of the perception. The process of cognition involves information processing and the content of mentality includes all kinds of information processed in the brain and the meaning of them. The information includes the input information from internal and external environment received by the brain and the output information generated by the brain to control body movement.

On aspects of cognition, there are sensation, perception, memory, attention, thinking and language, they all belong to mental activities. Sensation is the recognition of individual attributes of things generated in the brain as the objective things act on sensory organs. Perception is the overall understanding of the objective things inside the brain. Memory is the process of encoding, storing and retrieving the input information from the external environment or the information generated by the brain. Memory is a very important aspect of mental activity, including working memory which processes the current information and long-term memory. The long-term memory is further classified into episodic memory and semantic memory. Attention is the directivity and concentration of mental activity towards a certain target. Thinking is the process of analyzing, synthesizing, comparing, abstracting and summarizing information. Language is the symbolic system for human beings to communicate their thoughts, which is also a kind of mental activity.

Mind also has components of emotion and will. An individual's subjective experience on the aspects of emotions and will has psychological meaning. The emotions and feelings are the attitudinal experience of human beings towards objective things and corresponding reactions. The will is the psychological process of controlling and coordinating behaviors consciously to achieve predisposed goals through overcoming the difficulties, and the individual always has the will to his own activities.

Self-consciousness includes self-cognition, self-experience and self-control, etc. Self-cognition is the understanding of one's own self. Self-experience is internal experience accompanied with self-cognition. Self-control is expressed on behaviors.

From the preceding discussion, we can see the diversity and complexity of mind. All kinds of components of mind have close relations, and mental integration is achieved via the interactions among these components. In the previous chapter we introduced four functional systems of the brain; here the relation and interactions between various components of mind and these functional systems will be explained. The following excerpts are from the book *Unified Framework of Psychology and Cognitive Science* (Tang, 2007):

> The four functional systems of the brain are not isolated and unrelated; instead, there are interactions among them. Their respective functional activities and their mutual interaction and coordination form the integrated activities of the brain. The aforementioned four main components of mind, namely arousal and attention, cognition, emotion and will, and their mutual interactions constitute the integrated psychological activities. The four main components of psychological activities are closely related to the four functional systems of the brain.
>
> The four functional systems and their interactions lay the material foundation for components of psychological activities and their interactions. The arousal and attention are based on the activities of the first functional system; the cognition is mainly based on the

activities of the second functional system; the component of will is mainly based on the activities of the third functional system; and the emotion is based on the activities of the fourth functional system.

The interactions between the components of psychological activities have different brain mechanisms. The interaction between the component of arousal and attention with the other components is mainly realized by the interaction between the first functional system and the other functional systems. The first functional system ensures and coordinates the intensity and state of arousal, providing the basis for activities of the other functional systems, while the activities of the other functional systems will influence the first functional system.

The interactions between the components of cognition and the other psychological components are mainly realized by the interaction between the second functional system and the other functional systems. The second functional system receives, processes and stores information. The result of information processing will affect the activities of the other functional systems, while the activities of the other functional systems will influence the second functional system.

The interactions between the components of will and the other psychological components are mainly realized by the interaction between the third functional system and the other functional systems. The third functional system can generate behavioral procedures, and predict and implement actions. It can modulate and control the other functional systems, while the activities of the other functional systems will influence the third functional system.

The interactions between the components of emotion and the other psychological components are mainly realized by the interaction between the fourth functional system and the other functional systems. The fourth functional system has the functions of evaluating information and generating emotional experience.

The evaluation-emotion functional system interacts with the first functional system in that the latter provides the basis for activities of the fourth functional system, while the result of evaluating information by the fourth functional system and the emotion and reaction generated by that system will affect the activities of functional system responsible for modulating the intensity and state of arousal.

The evaluation-emotion functional system interacts with the second functional system in that the functional system of receiving, processing and storing information will provide materials for evaluation-emotion functional system, while in the process of receiving, processing and storing information, the evaluation is constantly going on. The evaluation involves the perception of objective events, the interpretation of the meaning of events, the retrieval of experience and the comparison of the information of events with the stored information. The result of evaluation and the emotional experience will affect the process of receiving, processing and storing information.

The evaluation-emotion functional system interacts with the third functional system in that the result of evaluating the evaluation-emotion functional system is the basis of the activities of making procedures, modulating and controlling for the third functional system; the evaluation-emotion functional system evaluates the meaning of information, selects the meaningful information for individuals, passes it to the functional system of making procedures, modulating and controlling, guides it to accomplish the task of modulation, and makes it modulate and control mental activities and behaviors to achieve the ultimate goals; while the third functional system influences the process of evaluation and further changes the emotional experience.

Every mental activity has many integration processes. Sensation and perception are used to exemplify as follows. The sensation and perception of human beings are basic psychological processes. The physical stimuli from the external environment act on human's sensory organs, transform them into the nerve impulses, and deliver them to the brain via the peripheral nervous system, activating the related brain regions. When the activation level reaches a certain value, the corresponding sensational experience will be generated. The physical stimulus from the external environment is a kind of objective physical event, the neural transmission and activation of brain regions are a kind of physiological activity, and the sensational experience is a kind of subjective psychological activity.

For individuals, various physical stimuli received by sensory organs are information to be processed inside the brain. Through the information processing and the conscious activities inside the brain, the relevant information is explained, including the meaning of the information. The information generates subjective experience, but they are quite different. The process of synthesizing all the information inside the brain is information integration, while the putting of various subjective experiences together into the overall subjective experience through information processing and conscious activities is psychological integration. The information integration is the basis of psychological integration, but the latter is different from the former, since the subjective experiences have psychological meaning.

It is mentioned in psychology that sensation is defined as the experience of concrete features of objective things, while perception is defined as the integrated features of objective things. In the processes of sensation and perception, there are psychological integrations. For example, look at a red ball on a platform moving towards a certain direction. An object has various attributes, such as shape, color, position and movement. These different attributes cause different subjective experiences, with the experience of the shape of a ball, the color in red, the position on the platform, the movement towards a certain direction, etc.

People perceive not isolated properties but the overall attribute, that is, people integrate the preceding various subjective experiences into an integrated subjective experience, such as the overall impression of a red ball on the platform moving along a certain direction. This phenomenon is called binding in psychology. The experiential binding means the binding of all sorts of experiences. The above example shows the binding of vision and perception.

Human beings have eyes, ears, noses, tongues and bodies. All these kinds of sensory organs receive external stimulus, passing neural impulses generated by different stimuli to the brain through sensory pathways. Vision is one kind of human senses. Besides visual sense there are other senses, such as auditory sense, tactile sense, olfactory sense and gustatory sense. In the previous example, if the ball produces various sounds, people will also perceive it. At this time, the human brain will not only integrate the stimulus from visual pathways, but also integrate the experience generated by visual stimulus with it through auditory sense. In daily life, when people receive various stimuli, they will integrate them. For instance, when people are watching an artistic performance, they see the actions and hear the music. Thus these experiences are combined, becoming a sort of perceptual binding across sensory pathways.

Binding is the most common psychological integration phenomenon.

Perceptual binding is one of the problems investigated most extensively in psychology. In fact, in many processes of psychological activities, there exists binding. Psychologists discuss not only binding of perceptions but also binding of working memory. These two kinds of binding interact constantly with each other.

In brain science, a lot of research has been conducted on the neural mechanism underlying feature binding. The information processing of neural system is accomplished first by different groups of neural cells detecting the features of external stimuli, termed as feature detection; and then all the features detected by different groups of neurons are integrated to achieve the binding of features (Treisman, Sykes, & Gelade, 1977; Treisman & Gelade, 1980). For example, the visual information of objects includes information of time and space and other superficial features. The visual system has parallel ventral pathway and dorsal pathway. They pass respective pieces of information to parietal lobe and temporal lobe to be integrated (Underleider & Mishkin, 1982).

Von der Malsburg (1981) proposed the hypothesis of synchronous activities to explain perceptual binding, believing that the neural activities inside the brain have oscillatory activities. For a target to be perceived, its various features are detected by different groups of neural cells, while the synchronization of oscillatory activities between neuron groups integrates the information processed by different groups of neurons, thus the binding of different features can be achieved to form unified perception. Eckhorn et al. (1988), Gray et al., (1989) and Singer and Gary (1995) respectively conducted experimental researches on the 40Hz oscillation. Their work provides some primary evidence for the hypothesis of synchronous activities.

The integration phenomena of language are other examples. Language is the distinctive feature of human beings and a complex mental phenomenon (Chomsky, 1957; Lü, 1979; de Saussure, 1980). As is known to all that humans comprehend and produce language, these processes include integrative action and integration process of various features.

The process of comprehending language goes like this: to process the acquired language materials and retrieve the previous knowledge at the same time, and to combine them and construct the meaning of these materials inside the brain. In the whole process, integration of different features is undergoing inside the brain. Thus language comprehension is an integration process.

The process of producing language goes like this: to sort out one's own thoughts, to determine the content to be expressed, to transform the thoughts into language materials and to generate language. In the whole process, the integration of different features is undergoing inside the brain. Thus language production is an integration process.

The content of psychological integration is colorful with various forms. In the coming chapter, we will discuss several aspects of the psychological integration phenomena.

Chapter 3 Neuro-integratics, Brain Integratics and Brainics

In the previous two chapters, we have discussed mainly integrated brain and integration of the brain. In this chapter, we will point out that in order to study deeply about the properties, laws and applications of neural systems and the brain, we need three new disciplines termed as neuro-integratics, brain integratics and brainics.

Neuro-integratics concerns about the properties, laws and applications of the integration phenomena of neural systems; brain integratics is about the properties, laws and applications of the integration phenomena of the brain; while brainics studies how to imitate the brain.

3.1 Neuro-integratics

In the previous two chapters, we have discussed the integration phenomenon at the system level of the brain. The neural system is composed of many levels of integrative actions and integration processes with different properties. Sherrington (1906) once systematically discussed various integrative actions of neural systems.

At the microscopic level, he discussed the integrative actions of synapse and neuron, believing that it is possible to understand the integrative action of the whole brain at the level of synapse and neuron. At a more complicated level of neural system, he also mentioned the integrative action of spinal cord, holding that the basic function of central nervous system lies in its integrative action.

The integrative action of synapse performs like the following: the synapse integrates many inputs and the soma generates an output. Sherrington (1906) stated that every synapse is a coordinating unit.

An example of the performance of integrative action of single neuron is listed as follows: a motor neuron integrates the excitatory input and inhibitory input and manipulates the behaviors. Sherrington (1906) claimed that a single neuron is the cellular base of the integrative action.

The performance of integrative action of the spinal cord can be described as follows: the spinal cord integrates the excitatory and inhibitory inputs and generates integrated responses. Sherrington (1906) also declared that reflex is one of the basic modes of action for central nervous system and also an entity of reflection of the integrative action of central nervous system; the reflex of spinal cord is the result of coordination of various inputs.

These discussions from Sherrington are all about the integration of signals at different levels of neural systems. He stated that the feature of the integrative actions of nervous system is that they are not realized by material transportation between cells but by the transfer of neural signals, that the integration of neural signals is an important aspect of neural integration, and that the integrative action in this process integrates different neural signals to generate the overall output and meanwhile integrate the animal body into a united whole.

It has to be pointed out that information integration is only one kind of integrative process of neural system. In neural system, there are not only lots of information integration, but also structural integration and functional integration, etc. For instance, the processes of neurons being integrated into clusters of neurons and neural circuits include both structural integration and functional integration. Thus, to do research on the integration phenomena of neural systems, besides information integration, both structural integration and functional integration should be taken into consideration.

To summarize, at various levels, each part of neural system is organized into a coordinating unity through interaction with each other and with the environment. At different levels of neural system, there are different integrative components, integrative actions and integration processes, which form entity of different levels. Various integrative actions exist at different levels of neural systems, including material integration, energy integration, structural integration, functional integration and information integration, etc.

On the basis of the experimental facts of various integrative actions and integration processes at different levels, a new discipline, neuro-integratics can be built, which concerns about the properties, laws and applications of the integration phenomena of neural systems. Its most important connotation is the integrative action and coordination of neural system.

3.2 Brain Integratics

Lots of integrative phenomena exist in brain regions of specific functions and the whole brain. The brain is an integrated entity. The various integrative components at different levels inside the brain interplay with each other, constitute integrative entity of different levels, forms and functions, and finally integrate into the whole coordinating brain with complex structures and functions, generating colorful mentality and various behaviors.

From the perspective of the whole brain, there are many functional systems, which are the integrative components of the brain. The interactions between these functional systems provide the integrative action of the brain. The body where the brain resides and the external environment provide the integrative environment for the brain. In the integrative process of the brain, the functional systems are organized into an integrated unity through the integrative action inside the brain and among the brain, body and environment. Therefore, the integration process of the brain is dynamic.

At different levels of the brain, there are various integration processes, such as material integration, energy integration, structural integration, functional integration, information integration, and psychology integration. On the basis of experimental facts of these integration processes, a new discipline, integratics, can be built to study the properties, laws and applications of the integrative phenomena inside the brain. To be more precise, this discipline can be called brain-mind integratics.

The brain is a part of neural system, thus neuro-integratics includes brain integratics, but they have different emphasis. Neuro-integratics stresses the research on integration phenomena and principles at the microscopic and mesoscopic levels of neural system; while brain integratics emphasizes the research on integration phenomena and principles at the macroscopic level.

Since the 19th century, in the history of neuroscience studies, some neuroscientists have disputed whether the functions of the brain are implemented in different areas or they are implemented in an integral whole. It has to be informed that the brain integratics we are discussing here is based on modern science, which is different from the theory of integral brain in the past.

In the middle of the 19th century, Flourens (1960) proposed the holistic theory of brain functions. In the late 1920s, Lashley (1929) proposed the integrationism of brain memory. These two theories are opposite from localizationism, and thus are the representatives of non-localization of brain function.

These two theories believe that all kinds of specific brain functions should attribute to the whole cerebral cortex. They are connected to all parts of cerebral the cortex instead of a specific part. These two theories were influential in the history of neuroscience. However, large amounts of experiments of functional brain imaging in modern brain science have proved their one-sidedness.

The brain integratics is totally different from these two theories. They have different understandings of the concept "integration": in brain integratics, integration means that the related parts inside the brain form into a unity through integrative actions; while in the aforementioned two theories, integration is the synonym of integral or non-localizationism of functions. To be precise, these two theories should be regarded as an isoenergetic theory of the brain instead of the integrative theory of the brain.

In the history of neuroscience research, there have been many discussions related to the integration phenomenon inside the brain. Besides Sherrington's (1906) research mentioned above, Luria (1973) proposed the concept of coordination of functional systems in the brain, Treisman et al. (1977; 1980) discussed the integration of attention, Crick (1984) investigated the integration of vision and perception, and Edelman and Tononi (2000) explored the integrative actions inside the brain. They are all related to the integration phenomena of the brain, but they are sparse, and not systematic. One mission of the brain integratics is to integrate various works related to integration phenomena in the brain and to construct a unified theory of the brain.

3.3 Brainics

Brain science highlights many issues, mainly including exploring, understanding, protecting, exploiting and imitating the brain (Tang et al., 2006).

Exploring the brain is the non-destructive measurement of the structure and function of the living brain in order to gain insight into the brain. Understanding the brain is to uncover the nature of the structure and function of the brain, and to understand the working mechanism of the brain and the laws of mind. Protecting the brain is to prevent and treat the disease of the brain based on the knowledge of the brain. Exploiting the brain is to develop the potential of the brain and to improve the quality of human beings. Imitating the brain is to develop highly intellectual computers or machines based on the knowledge about human brain, namely "machines copying the brain".

We thus try to propose a discipline, especially for conducting research on the imitation of the brain. In English, it is called brainics, consisting of "brain" and "ics", with the latter part meaning "have this property". Brainics is an interdiscipline of brain science, cognitive science, mathematics, information science, engineering and technology science and so on, which belongs to a branch of brain science.

In the natural environment, various biosystems are the targets to be imitated by human beings when they conduct researches on science and technology. We already have bionics. The experts in bionics observe many special features and skills owned by various biosystems, study their principles and mechanisms, and apply them to some fields of engineering and technology. They invented some methods, techniques and devices, which imitate the properties and principles of these creatures. Thus they can contribute to the innovation of science and technology.

Brainics is a part of bionics. The task of brainics is to do research on the structure and function of the brain, including the mind which is the advanced function of the brain, to find out some properties and principles of the human brain and mental activities, to provide new designing concepts and working mechanisms for engineering and technology, and thus to develop engineering and technology which imitates the human brain, such as intellectual technology and intellectual machines.

All kinds of integration phenomena that exist in neural systems, especially in the brain and mind are important for the study of brainics. Neuro-integratics and brain integratics provide theoretical foundations for brainics in terms of the research on the integration in neural systems and the brain. Brainics strives to imitate the properties and principles of the integrative phenomena of the brain and mind, so new ways, techniques and devices of imitating the brain can be developed.

Part II

General Integration Theory

In Part I, we have discussed integration phenomena in the brain and the integration theory of the brain. In this part, we will further explore integration phenomena in nature, technology and human society, analyzing all kinds of integrative actions and integration processes in these fields, summarizing all kinds of general concepts of integration phenomena and constructing the theories of general integration based on experimental facts.

In this part, in Chapter 4 we will investigate some integration phenomena in different fields; in Chapter 5 we will discuss and summarize some general concepts of these integration phenomena and illustrate the key points of general integration; and in Chapter 6 we will discuss the characteristics of general integration theory.

Chapter 4　Integration Phenomena in Different Fields

After investigating the integration phenomena in the brain at different levels, it is necessary to shift our attention to other fields. Actually, integration phenomena not only exists in the brain at different levels, but also widely in nature, technology and society.

In this chapter, we will exemplify integration phenomena in various fields of nature, technology, cross-disciplinary studies and society.

4.1 Integration Phenomena in Nature

In nature, whether in the physical world without lives or in the biological world, or in the spiritual world with brain activities as material basis, various integration processes can be found.

The physical world is hierarchical. It includes the world at microscopic, mesoscopic, macroscopic and cosmoscopic levels. In daily life, what people encounter is the macroscopic world. In the cosmoscopic physical world, there are all kinds of celestial bodies and space substances while in the microscopic physical world, there are quarks, electrons, nucleus, atoms and so on. The mesoscopic world is between the macroscopic physical world and the microscopic physical world, such as some atomic clusters which are mesoscopic substances.

There are many examples of integration phenomena at each level of the physical world. In the physical world, there are a variety of condensation phenomena. Atoms, molecules and electrons construct integrations of all kinds, forming condensed matters. According to the spatial scale, some belong to the mesoscopic world, while others belong to the macroscopic world. The mutual actions among atoms, molecules and electrons decide the internal activities, structures, physical properties and external attributes of these condensed matters.

In the cosmoscopic world, space substances undergo different kinds of integrations. For instance, the galaxy is constructed by stars and interstellar matters. The mutual actions, such as gravitational interaction and electromagnetic interaction between stars and interstellar matters, dominate the internal activities and structures and determine the external properties of galaxy.

The hierarchical structure of the biological world includes biological molecules, genes, subcellular structures, cells, multicellular organisms, organs, individual organisms, and the whole biological world. All kinds of integration processes can be seen at all these different levels.

In a single cell, there is integration of subcellular structures and elements inside the cell. There are material transportation and energy transfer between the living cell and its outer environment, which can be seen as an example of integration between a living cell and its outer environment.

Multicellular organism is composed of many single cells. There are the integrations between cells and the integrations between cells and their outer environment. From the perspective of the inside of biological organism, there are structural integration, functional integration, information integration, and so on.

Sherrington (1906) once discussed many aspects of integration inside animals. Examples include: the integration of single cells into an organ, and the integration of various organs into a unified individual animal, which belong to structural integration; the coordination of various glands through chemical interaction and the material transportation through blood cycling inside the animal body, which belong to functional integration; besides, through the transmission of neural signals inside the nervous system, the disperse organs are unified into a consistent individual animal, which belongs to information integration.

It has already been mentioned in Chapter 2 that on the aspects of human brain and mental activities, many different kinds of integration can be seen. In the process of interaction between individual and external environment, the input is realized by human sensation, while the output is realized by human movement. In Chapter 2 we have introduced the integration phenomena in the sensory processes, and in the following part we will introduce the integration phenomena in human movement.

Human locomotive organs can be dominated by mental activities and can fulfill all kinds of activities. The motor system and the neural system both participate in the movement. On the aspect of the motor system, there is the coordination between muscles and skeletons; in the neural system, there is the coordination of the spinal cord, brainstem, thalamus, motor areas of the brain cortex, cerebellum and basal ganglia, and so on.

Even with a simple action, the neural system and the motor system have to work together to monitor the motion and carry it out. As far as complex motions are concerned, more factors will be involved. The selection of motion category, motion direction and size, and the intricacy of controlling motions all need the involvement of planning, executing, controlling motions, and making the coordination of all kinds of functions of motions. This is the process of integration of human movements.

All above examples clarify that there exist integration phenomena in the physical, biological and spiritual world.

4.2 Integration Phenomena in Technology

In the field of technology, we can also find various kinds of integration phenomena.

Electronic devices are being widely used in modern technology, the most popular applications of which are integrated components and integrated circuits. The design and manufacture of electronic devices and circuits is the best example of the integration process.

In optical technology, the best example is integrative optical technology; in communication technology, the best example is Internet integrative technology; in computer manufacturing technology, the best example is the computer integrative manufacturing system, which is an intellectual manufacturing system integrated by computer technology, electronic technology, information technology, automatic control technology, mechanical technology and modern management technology.

In the following, the integration of experimental apparatus of nuclear physics and the integration of brain-machine interface in neural technology are introduced as examples of integration in technology and engineering.

To conduct experiments, we need an experimental apparatus and techniques. The experimental apparatus of nuclear physics includes all kinds of nuclear detectors and nuclear electronic instruments. In the experiments of nuclear physics, nuclear detectors are used to measure and record the nuclear radiation, while nuclear electronic instruments are used to analyze and handle the signals produced by nuclear detectors and collect the data of the experiments. The experimental apparatus of nuclear physics is integrated by nuclear detectors and electronic instruments. Detecting components and modules of various properties are integrated into various nuclear detectors, while different kinds of electronic circuits are integrated into various nuclear electronic instruments.

There are many kinds of electronic circuits, such as discriminator, amplifier, gate circuit, fan-in and fan-out circuit, coincidence circuit, anticoincidence circuit, pulse-amplitude analyzer and pulse-time analyzer, and so on. They are respectively made into standardized electronic circuits. For example, the discriminator in the nuclear electronic apparatus is a kind of circuit which has the function of discriminating the pulse amplitude. If the input amplitude of the pulse exceeds the predetermined value (discrimination threshold), there will be an output signal in the electronic circuit; if the input amplitude of the pulse does not reach the threshold, there will be no output signal. The discriminator is integrated by electronic components according to some rules. Another example is the amplifier, which has the function of amplifying the pulse amplitude. An input pulse signal will have greater amplitude after going through the amplifier; this amplification might be linear or non-linear. The other electronic circuits have their own functions. These circuits are all integrated by electronic components according to different rules.

All kinds of electronic circuits have standardized components, which can be integrated by logical design according to actual needs. Then they can be further integrated with various kinds of nuclear detectors into complex experimental apparatus of nuclear physics. The integrated complex apparatus can select, recognize, analyze and record the nuclear reaction events which play an important role in experiments of nuclear physics.

Another example of integration phenomenon in the field of technology is the integration of brain-machine interface in neural engineering. Neural engineering is a cross discipline between neural science and engineering technology. The brain-machine interface (BMI) is a new interface technology that uses computers to retrieve signals from electric activities from the brain, and makes use of these signals to communicate with the external environment, such as controlling the work of machine, and fulfilling people's intentions (Wolpaw et al., 2002; Wickelgren, 2003).

Usually, all kinds of sensors or integrated chips which build the pathway of communication between the brain and the external equipment can be used to connect with the brain. After receiving and transferring the brain signals, the mechanical device controlled by the computer will take place of human organs and act out the movements. According to different needs, various kinds of brain signals can be used to command machines to fulfill different kinds of movement, acting as the external environment so as to communicate or monitor, etc.

In some laboratories, this kind of technology is used to serve those who cannot move because of disease (e.g. paralysis). They control the related devices to collect emails, switch on and off the electronic equipment, and control the movement of a wheelchair, in this way to communicate with external environment or control the machinery to some degree.

The BMI technology involves different kinds of technology of many fields. Among them, to use an electrode to receive brain signals needs experimental technology of biology and psychology; to use computers to deal with signals needs information and computer technology; to use computers to drive the mechanical device needs computer technology and mechanical transmission device. Therefore, the brain-machine interface is achieved by the integration of these technologies.

The aforementioned examples show that there exist all kinds of integration phenomena in the field of engineering technology.

4.3 Integration Phenomena in Cross-disciplinary Studies

One important attribute of modern science and technology development is the mutual intersection, interaction and integration of different disciplines. In many specific scientific fields, depth and intricacy make the discipline highly differential and professional; and meanwhile, the research of each discipline needs the involvement of other related disciplines. Through the knowledge and technology integration across different disciplines, the cross-disciplinary studies are emerging, which result in many new marginal disciplines.

There are many examples of cross-disciplinary studies. To take modern biology and medical science as an example, their development is very fast, and the cross-disciplinary studies between them and other disciplines are very broad. The following is an example of medical physics which will be used to illustrate the development in this field.

Medical physics is the cross-discipline of medical science and physics, including wide contents, among which nuclear medicine and radiotherapy are two applications of nuclear physics and nuclear technology in medical science.

Nuclear medicine applies nuclear technology to clinical diagnosis to better serve the people. For instance, in diagnostic medicine by injecting small a dose of radionuclide or stable isotope labeled tracers inside the patient's body, and meanwhile using an imaging instrument which can measure the radiation of the radionuclide or stable isotope, some diseases can be diagnosed with the distribution and metabolism of the diagnostic medicine. The usually used FDG-PET technology with diagnostic medicine of deoxyglucose labeled with 18F nuclide plays an important role in diagnosing cancer. The field of nuclear medicine requires the knowledge and technology integration of nuclear physics, radiation chemistry, medical science, pharmacology and computer science.

The radiotherapy applies nuclear technology to clinical treatment to serve the people. For instance, we can use an accelerator to generate bundles of rays, focusing on cancer, killing cancer cells to achieve the purpose of treating illness. Modern proton or heavy ions treatment use an accelerator to generate bundles of protons or heavy ions. The range and intensity of exposure can be accurately modulated and can perform well on treating tumors. The field of radiotherapy needs the knowledge and technology integration of the fields of nuclear physics, accelerator physics, radiophysiology, medical science, computer science, and so on.

There are many integration phenomena in cross-disciplines. They are complex integrations, including knowledge and technology integration of different disciplines, as well as integration of research teams, resources and management.

4.4 Integration Phenomena in Human Society

In human society, there are also many different kinds of integration phenomena. In the following, social group, social service and social intelligence will be taken as examples.

An individual is a member of a family, school, group or society. An individual lives in such kinds of environment as family, school, group or society. They cannot live without social environment. Social integration manifests as the integration of individuals into a group, who work and live in a certain social environment, and conduct social interaction and activities, while influenced by history and culture.

Team organization can be used as an example of social integration. Members in the group have the same goal, collaborate with each other and serve the society. The whole team should exert each member's initiative and form a harmonious community. This is the integration of a team.

There are many kinds of integration phenomena in economy, culture and education. In each field, there are various integrations of respective activities.

One example of the integration of social service is medical and health services. Our country is developing the three-level health service network and the demographic and health science data sharing platform in rural areas. The construction of three-level health service network is the integration of medical and health service institutions while the construction of demographic and health science data sharing platform makes the rural population share the related scientific data, and helps raise the health level of the whole rural population, which is the integration of medical and health information resources.

Another example is the integration of social intelligence. Human intellectual activities bear the feature of sociality. Many intellectual activities in human society are realized by activities of collective which is integrated by many individuals. Thus social intellectual integration emerges.

Dai Ruwei (2009) studied social intellectual activities in detail. He pointed out that in modern society, human groups and machines represented by computers live together, that people live together with nature, and that various kinds of ideas collide, encourage each other and progress together. He also said, "The main part of innovation involves social man, human group and scientific team. It involves many disciplines and engineering fields, and principally relies on engineering innovation team."

In Chapter 4 we have offered some examples of integration phenomena in nature, technology, cross-disciplines and human society. Although there are not many examples demonstrated, it is clear that the integration phenomena are very common in different fields.

In nature, technology, cross-disciplines and human society, many superficial unrelated phenomena, such as condensation of atoms, reflex of animals, design of circuits, cross study of disciplines, collaboration of teams have some common features. From the perspective of integration, they are integration phenomena of different kinds and with different properties. For instance, the condensation of atoms is physical integration; the reflex of animals is neural integration; the circuit design is component integration; the cross study of disciplines is knowledge integration; and the teamwork is social integration, etc. Because they have different integration components, actions, environments and integration processes, they form various integration unities and show various integration properties. They belong to different fields and have their own rules respectively; however, they also share the common properties of integration phenomena.

The viewpoint of integration is used to investigate various integration phenomena in different fields which helps us summarize their general concepts and properties, know their general principles and laws, apply these concepts and laws to research on complex things, and deal with complex events related with integration phenomena.

Chapter 5 Exploring General Integration Theory

In the previous chapters, we have introduced the integrated brain, the integration of brain, and the integration phenomena in different fields, suggesting that integration phenomena widely exist in nature, technology and human society.

In this chapter, we will propose the general integration theory based on many experimental facts. We will first discuss the common properties of integration phenomena and illustrate the main points of general integration theory, and then summarize some general concepts of all kinds of integration phenomena.

5.1 General Integration Theory

Integration is a process of constructing a unity with new functions by integrative components and their mutual interactions. Our research on integrative phenomena is learning from the brain. And these integration processes exist inside the brain, and widely in nature, technology and society.

As mentioned previously, in nature, there are lots of integrative actions and integration processes at different levels. In nature, including the physical world, the biological world and the spiritual world, all kinds of things constitute various integrated unities at different levels, and they respectively have different attributes. From the perspective of the spiritual world, there are many different integration processes in mental activities, and even human consciousness is produced in the process of integration of brain functions. In human social activities, there are all kinds of integration processes. The individuals interact with each other and thus constitute collective and society.

Different integration components and their mutual interactions have their own properties. Different kinds of integration processes have their unique properties and rules, which need specific analysis and research. As far as general integration processes are concerned, they have common properties and concepts. General integration theory can be applied to investigate different integrative actions and integration processes in different fields, find out their common properties and rules through comprehensive research, and discuss their application in specific fields based on these general properties and rules.

These cocepts of integration have existed for many years. We are familiar with the term 'integration' and use it on different occasions, such as integrated circuit, and integrated container, etc. However, in general integration theory, the term 'integration' is more meaningful.

What are the common features of integration phenomena? Specialized research is needed for this question. Our task is to collect all kinds of integration phenomena in nature, technology and human society together, make integration phenomena as a specialized subject to study, build a new discipline and conduct research on them.

General integration theory points out that integration phenomena are common in complex systems. In the process of integration, many integration components form into a harmonious unity through mutual interaction and their interaction with the environment.

Integration is a dynamic process. The integrated unity is a whole one. Many components inside the unity are called integration components; the mutual interaction inside the unity is called integrative actions; the environment where integration processes happen is called integration environment; the process of integration components forming into a unity is called integrative process; the product of this process is called integrated unity.

The integrative process usually includes many integration components. All kinds of integration components and their mutual interactions are the basis of integrative process. These components are small elements participating in the integration processes and composing the integrated unity. The complex system is composed of various components. Some complex systems have hierarchical structures. At each level, there are different kinds of integrative actions, integration processes and integrated unity.

There are many kinds of integration components. It has been mentioned previously that in the physical and biological world, integration components include material, energy, structure, function and information, and thus the integration processes include the integration of all these components. In the spiritual world and human society, there are other integration components.

In daily life, people tend to understand more of structural integration, such as integrated circuit and integrated container. General integration theory not only conducts research on the structural integration, but also conducts research on integration of material, energy and information.

As mentioned above, learning from the brain provides profound materials for general integration processes. Research on the brain and mind is not only about its structure and function, but also physiology, psychology and pathology, including subjective experience, cognition, emotion, intention, consciousness and behavior, thus the integration processes of the brain include the integration of material, energy, structure, function, information, psychology, behavior, and the brain, mind and society.

Integration is not simply a collection of components. It must be based on the interaction between the components, which means that components without interaction cannot be integrated. The whole process is going on in a certain environment. The integrative components inside the system interact with each other and with external environment to form a unity.

Integration is a developing process, with lots of integrative components constructing a new unity with new functions in this dynamic developing process. We can use some parameters to describe the properties of integration processes, such as the degree of integration and the speed of integration, etc.

To take a neural system as an example, Tononi et al. (1994) once discussed the degree of integration of neural system $I(X)$. The system X consists of several (n) parts of x_i, each individual part has the entropy $H(x_i)$, and the system as a whole has the entropy $H(X)$.

They define the degree of integration of system X as the sum of $H(x_i)$ minus $H(X)$, namely:

$$I(X) = \sum_{i=1}^{n} H(x^2) - H(X)$$

In the equation $I(X)$ shows the decrease of entropy because of the interaction between these parts. The stronger the interaction, the bigger the value of $I(X)$.

In the integrative process, the higher degree of integration increases, the more new properties that the previous components do not possess will show up. This phenomenon is called emergence.

The integrative process is realized by the interactions among the integration components. The integration components and their interaction have diversity, and thus there are various types and forms of integration processes with their own properties. Various kinds of modules and networks which are formed in these processes generate an integrated unity. In the unity, every component coordinates with each other.

Since there are many kinds of integrative phenomena in nature, technology and human society, we think it is necessary to build a discipline that we term as general integration theory to be specially applied to the research on integrative phenomena. Therefore, general integration theory is a discipline which studies the general properties, rules and applications of integrative phenomena in nature, technology and human society. It not only studies the general properties and rules of integrative actions and integration processes, but also discusses how to effectively integrate the components and how to innovate by the integration based on the properties of matters.

Von Bertalanffy (1950; 1976) referred to his system theory as general system theory, for what he studied is not some specific system, but a general system that exists in nature and human society. Similarly, in general integration theory, what we discuss is not some specific integrative phenomena, but general integrative phenomena in nature, technology and human society. Therefore, we term it as general integration theory.

As a new discipline, general integration theory has certain research objects, research goals, research contents and core concepts.

First of all, it has research objects. It studies the integration phenomena at different levels with different properties in nature, technology and human society. From the facts of integrative phenomena, their common properties can be summarized.

Secondly, the goals of general integration theory are to build a new discipline to study general properties and rules of all kinds of integrative phenomena, and to apply it to the fields of nature, science and technology and human society. It also studies the properties and rules in many specific fields and attempts to build a group of new disciplines called special integration theory, or special integratics.

Thirdly, the contents of general integration theory mainly include the common properties and concepts of all kinds of integrative phenomena at different levels in different fields, and collect these common things together to conduct a comprehensive research.

Fourthly, the core concept of general integration theory is integration. For each integrative phenomenon, the integration components, integrative actions, integration processes and the integrated unity all need to be investigated. Material integration, energy integration, structural integration, functional integration, information integration, mental integration, knowledge integration, environment integration and social integration all should be discussed. Some common concepts should also be summarized, such as global, globalization, module, modularity, reduction, reasonable reduction, synthesis, organic integration, binding, association, construction, optimization, criticality, emergence, complementarity, coordination, coincidence, synchronization, harmony, fluency, adaptability, assimilation, metasynthesis and grand unification. These concepts will be discussed in later sections.

It should be reminded here that general integration theory is different from the set theory in mathematics. Set theory is a branch of mathematics (Zehna & Johnson, 1986; Fang, 1982). In set theory, the whole collection of something that has the same certain property and can be differentiated is called a set. This branch of mathematics does not consider the special properties of the members in the set, but studies the properties of the set as a whole.

The concept of "set" in set theory is different from the concept of integration in general integration theory. In set theory, "set" is a math concept, emphasizing the collection of elements while in general integration theory, integration refers to all kinds of integrative phenomena in nature, technology and human society, especially integrative actions and integration processes. However, the two concepts are somehow connected because the integrated unity includes all the integration components.

The set theory and general integration theory are different theories. Set theory studies the mathematical properties of "set" while general integration theory discusses integration phenomena and their rules in the fields of nature, technology and human society, focusing on the properties of integrative actions and integration processes. Of course, in the research of general integration theory, some related mathematical tools in set theory can be used.

General integration theory provides us with theories about the general rules of integration phenomena, giving us not only a new kind of view about the world and matters, but also a method of dealing with things and solving problems.

Integration is not only a general principle, but also a viewpoint of observing and studying objects. Since integration phenomena exist widely in nature, technology and human society, it is reasonable to use the viewpoint of integration to observe and conduct research on all kinds of complex things which include complex integration phenomena.

For complex things, it is necessary to investigate their integration phenomena from many aspects, especially integration components, integrative actions, integration processes and the integrated unity. For example, if we study a kind of complex organism, not only its internal material integration, energy integration, structural integration, functional integration and information integration need observing, but also its integration with other organisms and environment need paying attention to.

Structural integration, functional integration and information integration are all complex processes. For a specific integrative process, we should investigate which integration components are involved, what interactions they have, and what are the properties of the interactions and the internal mechanism, etc. Here many concepts such as binding and construction are involved.

An integrative process is a dynamic process. For a specific integrative process, the related time properties and integrated dynamics should be considered. For example, what are the time properties of the integrative process, how the organizational structure of integrated unity evolves, and how new functions emerge in the integrative process, etc. Here the concepts like synchronization and emergence are involved.

Complex integrated entities are all integrated by their internal integration components. How these components interact and coordinate should be considered. Here the concepts like complementarity and coordination are needed.

Integration is not only a view of observing and studying the world, but also a way of handling and solving problems. Since integration phenomena exist widely in nature, technology and human society, it is reasonable to use the viewpoint of integration to observe all kinds of complex things which include complex integration phenomena and conduct research on them.

Integration is a method of constructing a unity with scattered components. An issue of achieving this is how to effectively integrate these components with different properties. In many integration processes, we usually construct modules and networks of different levels and properties according to the overall goals, then generate the unified output or products. Here concepts like modulization, globalization and optimization are needed.

When we research complex things, we always need to consider how to analyze and divide them into small parts and make connections and synthesis. The method provided by general integration theory is reasonable reduction and organic synthesis, namely reasonably reduce the complex thing to each part respectively and study them in detail, then integrate them organically based on their connections and interactions. Here, the concepts of reduction, rational reduction, synthesis and organic integration are involved.

To deal with complex things or solve complex problems, it is necessary first of all to analyze and discuss the related information, then to evaluate the situation and make decisions, and finally to organize the work and implement the plan. In all these processes, the method of general integration theory can be used. For example, in the process of analyzing the related information, we should understand the whole situation, integrate all the information and get accurate knowledge; in the process of making a decision, we should widely collect views, and integrate them to form an appropriate scheme; in the work of organization, we should reasonably distribute the resources, and should integrate all departments to coordinate a team; and in the process of implementation, we should integrate all processes to achieve perfect results under the unified command.

For those things containing integration processes, exploiting the viewpoint and method of general integration theory can help us understand and solve many real problems related with integration processes. Of course, in different fields, each integrative process is different, and thus we need to conduct specific researches on each integrative process. In the forthcoming Part III, we will discuss some applications of the theory in several specific fields.

5.2 Globalization and Modularity

Global and globalization are the concepts related with integration phenomena. Global refers to the whole situation while globalization means to coordinate the whole situation.

In the integrative process, first of all, the goal of integration should be based upon the whole situation; then the whole situation should be coordinated according to the full-scale integrative design; only in this way the final integrative process will be accomplished. Globalization is not only a concept of structural integration, but also an important concept of functional integration and information integration.

Psychologists understand concept of global workspace when they discuss models of consciousness (Baars, 1983; Baars & Franklin, 2003). The global workspace model believes that in the brain there exist many specialized processors and a global workspace, and that specialized processors deal with all kinds of information and can work independently while global workspace receives the information from all the specialized processors. As long as the information is expressed in the global workspace, it can be visited by rational behaviors and generate consciousness. Global workspace can broadcast the content of consciousness to neural systems of the brain and integrate the dispersed and independent brain functions. We have extended on the model of global workspace based on experimental facts and proposed the extended theory of global workspace of consciousness (Song &

Tang, 2008).

Another set of concepts related with integration phenomena is module and modularity. Module refers to standard component while modularity refers to the making of standard components in structural integration and then the construction of these components into a whole.

When constructing complex integrative unity, we can divide the integration processes into several steps according to the global goals. All kinds of modules are first of all constructed and then they are integrated into a complex integrated whole.

In cognitive neuroscience, there are theories about modules in the brain (Matthei & Fodor, 1984), which believe that the brain is constituted by highly specialized and independent modules. These modules are complex and intricate, whose combination lays the foundation of complex and intricate cognitive functions.

In psychology, when discussing long term memory, there is a concept of chunk. The span of short term memory is very limited, with only 7±2 items. However, we can use previous knowledge and experience to integrate information into chunks. Through enlarging the information volume of a chunk, the overall information of short-term memory can be increased.

The concept of module can be applied to structural integration. It is also related to functional integration and information integration, and can be applied to the integration in engineering and technology, team integration and social integration.

5.3 Reduction and Synthesis

In the integrative process, many integration components form an integrated unity through integrative actions. Reduction and synthesis are two concepts related with integration phenomena.

Reduction means analyzing the integration components in the integrated unity. Synthesis means integrating integration components into a unity. Reduction and synthesis are indispensable to the integrative process.

In section 5.1, the method of reasonable reduction and organic synthesis has already been mentioned. In the book *On Consciousness—Natural Science Research on Consciousness* (Tang, 2004), this method has been discussed to study complex matters. Let us review the method as follows:

> First of all, the method of reduction can be used to study consciousness. It is composed of many elements. The method of reduction is to break it down into elements and their interactions and conduct specific research on them. In this sense, consciousness should and can be reduced.
>
> Secondly, this kind of reduction should be moderate. Its result should have psychological meaning. We have analyzed the basic elements of consciousness and their interactions. Because they are basic and have psychological meaning, this kind of reduction is reasonable. The research on consciousness does not have to be at the level of a single ion channel of a neuron.
>
> To use the method of reasonable reduction to study consciousness is just one aspect of studying consciousness. Another aspect of studying consciousness is to use the method of organic synthesis. It might be beneficial to combine these two methods.
>
> The method of organic synthesis is based on reasonable reduction. We should further understand the organic relations between the reduced elements, then combine these elements

organically to achieve the understanding of complex things.

5.4 Binding and Association

Binding is another concept related to integrative phenomena. In Chapter 2 we have already mentioned that binding is an important concept of perception in psychology, which means bundling together.

Pattern recognition is an example. A figure has features of many aspects, including dots, lines, angles, directions and movements. In the process of pattern recognition, the visual system checks many independent features of the figure. However, in fact, a visual system can combine all the features of the figure together to achieve the overall perception of the figure. How does the brain achieve this? This is called the feature binding, which also belongs to perceptual integration.

Currently, people believe that attention mechanism is the key to perceptual integration. If there is no attention, the features are separate; when one's attention participates in the activity, the brain can bind all the features, thus perceive the whole object (Treisman, Sykes, & Gelade, 1977; Treisma & Gelade, 1980). This is the feature binding in perception. The meaning of binding can be expanded. For example, when we refer to information integration and knowledge integration, binding means putting different information and knowledge together.

Another set of concepts is union and association. Union means connection and combination, and association means joint of concepts.

In the components of a union, there is always an aspect of antagonism. For example, the inputs of neural system may be excitatory or inhibitory. The integrative action of neural system is to integrate all these inputs and produce the overall behavior.

In psychology, association and binding are two ways of cognitive processes inside the brain. The image of things in the brain is called representation; association is to associate one thing with another, integrate the former representations to produce new representation; the binding is to bind every feature of the same thing to produce a new representation.

The brain can integrate all kinds of features of the detected things to form a certain cognitive structure, and it also makes explanations and predictions based on the memory.

Association is a way of conceptual integration. It can reach other concepts from one concept, or collect different concepts to form complex concepts, or find out the relations between different concepts. Association is not only important in psychological integration, but also an efficient way of knowledge integration.

5.5 Reconstruction and Optimization

Reconstruction and optimization is another pair of important concept of integration phenomena. The integration process is the process of continuous construction and reconstruction.

The integration process is complex. In the integration process, it is always necessary to discard the dross, select the essential and reconstruct the structure. For instance, during the development of neural system, the useless part will be deleted, which is called the pruning of neurons.

Biological evolution is another example. All the living creatures are evolving and the existence of current creatures result from natural selection. Jacob (1977) once referred to evolution as a tinker.

In the process of constructing theories, it is necessary for us to have the viewpoints and methods of reconstruction. Crick (1994) once talked about the experience of constructing biological theories in the book *What Mad Pursuit: A Personal View of Scientific Discovery*. He said: "You may think that you can produce a useful theory just by connecting a witty idea and imaged facts, but that will be unreliable. Those who think they can think out a new theory in the first place are also wrong. Before the big success, they should propose one theory after another. It is the process of constantly giving up a theory and adopting another theory that makes them critical and just, which is necessary for their success."

In the following, another important concept 'optimization' should be explained. Optimization means making something more complete and close to be perfect.

In the integration process, there is an optimization process of integration goals, integration components, integrative plans and methods, etc. The optimization of integration goals is to produce the best integration as the goal; the optimization of integration components is to select better components to complete structural integration; the optimization of integration plan is to design the best plan to fulfill structural and functional integration; and the optimization of integration method is to choose the best methods or ways to complete the integrative processes, etc. For those integration processes in nature, there exist natural selections. For those integration processes conducted by people, there are artificial selections. It is always necessary to evaluate, examine, compare and choose to achieve the optimization.

The concept of optimization can not only be applied to structural and functional integration, but also is important to other kinds of integration processes. People conduct integrations with the requirement of optimization and efficacy. The optimal integration is also effective integration.

5.6 Criticality and Emergence

In the integrative process, the system undergoes quantitative change. When the quantitative change accumulates to some degree, the system will have qualitative change and emerge new features under the critical conditions. Emergence refers to the phenomenon that under certain conditions, in the integrative process, the system undergoes qualitative change and new features emerge. The critical conditions are the conditions of qualitative change for the emergence of new features of the system.

In the book *On Consciousness—Natural Science Research on Consciousness* (Tang, 2004) the critical conditions are explained as follows:

> In daily life, we often see phase transition in physics. In physics, phase is used to describe the state of matter. The same matter has different states, such as gas, liquid and solid state. These are different phases.
>
> Different phases are different states of matter. The matter transfers from one phase to another, which is called phase transition. When the matter does not reach the critical conditions, it keeps one kind of phase; when it does, it transfers from one phase to another, for example, from gas state to liquid state, or from liquid state to solid state.
>
> An example of phase change in the non-equilibrium system of physics is: the gas in an expansion cloud chamber condenses and changes into drops of water along the track of electric particles.

Another example is that the overheated liquid in the bubble chamber transfers into bubbles along the track of electric particles, which is the phase transition from liquid state to gas state.

In that book, the emergence of consciousness is discussed as follows:

> Emergence of consciousness is an important phenomenon of brain activities. When the activation level of a certain brain region does not reach the critical conditions of emergence of consciousness, the information processing in this brain region is unconscious; when it does, the information processing will suddenly transfer from unconsciousness to consciousness. When consciousness emerges, subjective experience generates corresponding to this activated brain region.
>
> The emergence of consciousness is sudden, but not continual. Activities in a certain brain region enter consciousness or not without ambiguity. The brain is an open and non-equilibrium system, and the emergence of consciousness in the brain is similar to the symmetry breaking in physics.
>
> Many phenomena happen in the process of emergence of consciousness, including: the activation of given brain regions and other brain regions; the competition of attention resource between different brain regions; the increase of activation level of given brain regions due to attention and the access of information processing to consciousness when the activation level of given brain regions reaches the threshold of consciousness. There is a close relation between these phenomena. The process of consciousness emergence is a dynamic process of resource competition, selection and elimination.

5.7 Complementarity and Coordination

Complementarity means to complement each other.

The concept of complementarity has existed since ancient times in China. For example, the concept of interdependence between Yin and Yang and the unity of heaven and human.

In physics, Bohr (1928) proposed the principle of complementarity. This principle points out that light and matter both have wave-particle dualism. To describe microscopic physical phenomena, waves and particles are both indispensable. These two concepts are coordinate, or in a sense, complementary.

The principle of complementarity brings to light that in describing microscopic physical phenomena, some classical concepts reject some other concepts, while both are necessary for comprehensive description of microscopic phenomena, and that only by combining these exclusive but complementary concepts, can the comprehensive description be achieved. To give a common example, a coin has two sides which are both necessary. Only by seeing both sides the overall knowledge of it can be gained (Bohr, 1964).

In the previous discussion, we mainly talk about different concepts of complementarity. In general integration theory, we expand it to include the complementarity of different parts in functional integration or knowledge integration, for example, the complementarity of resource advantages or the complementarity of theoretical viewpoints.

When we propose the unified research approach of psychology, we try to point out that in modern psychology, there are different research approaches. They do not have contradictory viewpoints, but their concerned issues, focuses, and research methods are different. We believe that some concepts of a research approach are complementary, thus the unitary research approach of psychology is possible (Tang, 2007).

In the integrative process, we should effectively integrate different sides to make the system work better.

Another concept is coordination which means cooperating appropriately and being consistent.

It has been mentioned before that the overall function of the brain is realized by several functional systems in the brain. These brain functional systems have respective functions and also integration. The mind and behavior are the results of the coordination of several functional systems in the brain. This is a good example of function coordination. These brain functional systems coordinate properly and harmoniously to make sure normal mentality and behavior. In opposition, if the brain functional systems work disorderly, one's mind and behavior will be abnormal, which needs adjustment (Buzsaki, 2006).

Coordination is not only a concept of functional integration, but also a concept related to engineering integration, technology integration, management integration and social integration, for example, the efficient work of lots of machines in engineering integration and the construction of harmonious communities in social integration.

5.8 Coincidence and Synchronization

The meaning of coincidence is to make one in accordance with another.

Coincidence register is an electronic device of selecting and recording the synchronic signals. The corresponding circuit is called coincidence circuit. For example, double coincidence circuit is a kind of electronic circuit with two channels and coincidence functions. Only when two input signals arrive within a short time interval, there is coincidence output. There is also multiple coincidence circuit with multiple channels.

One relevant device is anti-coincidence circuit. When there is an input signal from coincidence channel and there is no input signal from anti-coincidence channel, there is output pulse; but if there is an input signal from anti-coincidence channel, there is no output pulse. Another example is delayed coincidence circuit. Only when there is one input signal from one channel, and another input signal from another channel after a given time interval, there is an output pulse of delayed coincidence.

In information integration, coincidence refers to the method of choosing signals at a proper time in the integrative process. To use this method we can record the signals that happen at the same time or select the events that happen in different places at the same time. To summarize, this method can explain how to choose the common features of different events, which is useful in information integration and knowledge integration.

Let's study another concept—synchronization, which means changing simultaneously.

Synchronized acceleration is a kind of principle and method of accelerating the particles in high-energy accelerator. Synchrotron is a large experimental device used to accelerate high-energy particles. Synchron radiation facility is also a large experimental device to use electron synchrotron to produce optic radiation.

In high-energy physical experiment, the principle of synchronized accelerating particles is to be used in the high-frequency electric field to accelerate electric particles. As the particles speed up and the mass of the particles increase due to relativistic effects, the intensity of magnetic field of the accelerator increases, and thus the electric particles can be accelerated in the circular orbit with constant radius.

Synchrotron radiation light is the light radiation produced when electrons do curve motions in the magnetic field of electron synchrotron, including visible light and light in X-ray wave band. This kind of large-scale experimental device is called a synchrotron radiation light source device.

Synchronization also refers to the change of related functions in functional integration and information integration together. This concept can also be extended to mean the coordination and interaction of each part in these integrations. Synchronization ensures the harmony and stability of the system.

Another concept of integration is fluency. Fluency means that the material transportation, energy transfer and information transmission are fluent.

It has been mentioned that in the brain, the information processing has several levels, including upper levels and lower levels. There is information flow between these levels. The information processing is the interaction between the information flow from bottom to top levels and that from top to bottom levels.

In the information processing, the information flows through information pathway, thus the fluency of the pathway is an important condition of information processing. When an activated brain region acts on another region, their interaction will make the pathways more and more efficient. This is called facilitation of the pathways (Tang, 2003a).

Fluency used to refer to the smooth flow of information during information integration, but now this concept can be extended to effective connections in structural integration and functional integration.

5.9 Adaptability and Assimilation

Adaptability is concerned with environment integration, which means that everything should adapt to the environment.

Each kind of thing is not independent, but is circled by its environment. Environment is always changing, thus things must make adjustments to fit the changing environment. The effective integration of things and environment requires that they work friendly to each other to form an effective and interactive unity.

In cognitive science, there once was a concept of 'situated cognition' (Brooks, 1991). It believes that cognition is in situations and relies on situations and that cognitive activities and situations are closely related and cannot be separated.

Adaptability is not only a concept of environment integration, but also a concept of biological integration and social integration.

Another two concepts concerned with integration are assimilation and accommodation. They were proposed by Piaget (1983) when he studied cognitive structures. In psychology, these two concepts refer to both ways of cognitive adaptation. Assimilation is putting new information into established cognitive structure, while accommodation is changing the pre-existing cognitive structure to adapt to new environment and information.

Piaget's genetic epistemology states that people interact with environment in the process of understanding the world to form their own cognitive structure, which is called schema. One kind of schema undergoes assimilation and accommodation to become a new kind of schema. In this way, individuals constantly adapt to the changing environment through interacting with the environment.

These two concepts can also be used in theory integration and social integration to discuss adaptability.

5.10 Global Integration and Grand Unification

Concepts related to large-scale integration are global integration and grand unification.

In China, we have held the concepts of global integration and grand unification since ancient times. From the perspective of integrative process, global integration is a large-scale integrative process, integrating all the components into a grand unification. Thus the grand unification is the result of global integration.

In constructing ideological system, we need the concept of global integration and grand unification. *Unified Framework of Psychology and Cognitive Science* (Tang, 2007) uses these two concepts to discuss the theory of grand unification of psychological interaction and the unified research approach in psychology.

In expounding the theory of grand unification of psychological interaction, the following is pointed out in this book:

> This theory includes all kinds of psychological interactions, both discriminating their different features and pointing out their common and unified basis. This theory is a unified theory, covering all the psychological interactions, thus it is called grand unification theory.

In explaining the unified research approach in psychology, the following is pointed out in this book:

A kind of psychological phenomenon always involves many kinds of psychological interactions. Since these interactions are not independent, we should use the concepts of connections and unifications of multiple psychological interactions to investigate this kind of phenomenon.

The unified research approach in psychology comprehensively studies all kinds of psychological interactions and emphasizes the unification of these interactions. Thus its research fields include all the psychological fields, and its research approach is wider than current research approaches. This approach can unify the primary and reasonable viewpoints of different research approaches, based on which, the advantages of different research approaches can be integrated.

Chapter 6 Characteristics of General Integration Theory

In Part I we have investigated experimental facts the about brain, and proposed the integration theory of the brain. In the previous two chapters of Part II, we started from integration phenomena of the brain and reached the research of general integration theory.

This chapter illustrates the characteristics of general integration theory, emphasizes that general integration theory is based on learning from the brain, and proposes that the theories of general integration theory can be applied to all kinds of specific fields, constructing all kinds of theories of special integration and illuminating the connections and differentiations between general integration theory and system theory.

6.1 General Integration Theory Based on Learning From the Brain

In the 1980s, some scholars proposed that there are two ways of studying the science of thinking: one is to study the brain and figure out the mechanism of brain activities; the other is to find out human's thinking pattern through artificial intelligence and use computers to imitate human brains, combining the science of thinking with artificial intelligence and intelligence machines. At that time, brain science was thought to be complicated, so artificial intelligence was chosen as the research approach (Qian, 1986).

Now the situation is different. As the experimental technology of brain science develops quickly to provide all kinds of conditions for substantial research on human thinking, the way of studying brain mechanism is possible (Tang, 2003a). We believe that mind is the advanced function of the brain, and that we have to understand our brain in order to understand our mind. Thus the science of thinking should combine these two ways of studying and thinking, and meanwhile emphasize the basic position of brain research, and conduct theoretical and experimental research on the brain as well (Tang et al., 2006).

The notion should not only be applied to the research of science of thinking, but also to the research of general integration theory. To study general integration theory, we should begin from the brain, investigate the integration phenomena of the brain and learn the integrative principles of the brain. The brain is the most complex system in nature, the activities of which are the most complex ones, and the structure and activities of which can provide us with profound experimental materials about integration phenomena.

One feature of general integration theory is to learn from the brain. General integration theory firstly studies the integration phenomena, and then develops the general integration theory on the basis of learning from the brain. There are many kinds of integrative actions and integration processes in the brain. To study them is beneficial for developing the concepts and theories of general integration theory.

When we research natural phenomena, including the physical

world without life and the biological world full of life, we usually discuss the motion of matter, energy transfer, structural connection, function regulation and information exchange and so on, without involving subjective mental activities, if we do not consider the activities of the brain and mind. In these discussions, we include the concepts of matter, energy, structure, function and information.

To learn from the brain makes us know more widely and abundantly about the world. When we study the structure and activities of the brain, we not only face the physical brain and biological brain, including the physical phenomena and biological phenomena, but also mentality and behavior, namely the interactions between the spiritual world, mentality, brain, body and the natural and social environment.

The brain and mental phenomena involve the physical, biological and mental world. When we investigate the activities of the brain and mentality, we should discuss not only all kinds of motions of matter, energy transfer, structural connection, function regulation and information exchange, but also colorful subjective experience, such as cognitive, emotional and volitional activities, and relevant social activities. Thus, the concepts involved in brain and mental phenomena include not only the common concepts with life activities, but also other cognitive concepts, such as experience, cognition, emotion, volition, consciousness and behavior.

The general integration theory proposed in this book begins with brain science, based on which we develop the general integration theory. In studying the integration phenomena in the brain, besides studying material integration, energy integration, structural integration, functional integration, information integration, we should also study subjective experience related with mental activities, such as cognition, emotion, volition and human behavior, which provide us with a lot of examples of integration processes in mental, behavioral and social fields.

Neuro-integratics, brain integratics and psycho-integratcs are all theories in the fields of the brain and mentality, which belong to specialized theories of brain science and psychology. Our purpose is not only to build neuro-integratics, brain integratics and psycho-integratics, but also to further develop general integration theory.

To achieve this, we should learn from the brain, investigating all kinds of integrative actions and integration processes in nature, technology and human society. General integration theory is to study all kinds of integrative actions and integration processes in different fields so as to find out general features and common rules.

6.2 General Integration Theory and Special Integration Theory

As mentioned in the last section, general integration theory begins from the research of brain science and further investigates integration phenomena in different fields. Chapter 4 has already given some examples that show integration phenomena widely exist in different fields. General integration theory has the research range of all kinds of integration phenomena in nature, technology and human society.

The previous section has discussed one feature of general integration theory, namely learning from the brain. This section will illustrate another one, that is, the common features of different kinds of integration phenomena.

As a discipline, general integration theory not only discusses lots of integration phenomena in some specialized fields, but also exemplifies all kinds of integrative actions and integration processes in different fields, and then studies the common features of them. What it discusses is not only the relevant concepts of some specialized integration processes, but also the common features and general concepts of all kinds of integration processes in different fields.

We have studied the general features of all kinds of integration phenomena, summarized general concepts of them and illuminated the general views about them. We term this new discipline as general integration theory. Based on the research of general integration theory, we should also use the view of general integration theory to study all kinds of integrations, investigate their features, rules and applications of integrative actions and integration processes in different fields. To apply general integration theory to all kinds of specialized fields, we have to develop a series of sub-disciplines to study integration phenomena in different fields. We call these sub-disciplines as special integration theory or special integratics.

General integration theory and special integration theory have similarities and differences. General integration theory discusses all kinds of integrative actions and integration processes in different fields, which is the integration in a broad sense while special integration theory discusses the features of integrative actions and integration processes in a specialized field, which is specific and narrow. In principle, general integration theory includes the common features and rules of all the special integration theory.

In the past, there were scholars discussing integration in a specific field and putting forward interesting viewpoints. However, the present book is different from them. It mainly focuses on general integration theory, namely the general features of all kinds of integration phenomena in nature, technology and human society. This book puts emphasis on the integrative actions and integration processes in the brain to provide basis for developing general integration theory.

Special integration theory is to apply general integration theory to a specific field, to study the features of integration phenomena in that particular field and the specific rules and applications.

In nature, the examples of special integration theory include: biological integratics which studies the features, rules and applications of integration phenomena in biology, cell integratics which studies those in cells, neuro-integratics which studies those in neural system, brain integratics which studies those in the brain, and also human body integratics, med-integratics, psycho-integratics, cogno-integratics, intelligence integratics, environment integratics, geo-integratics, space integratics and so on.

In technology, examples of special integratics include info-integratics, engineering integratics, technology integratics, etc.

In humanities, examples of special integratics include culture integratics, art integratics, education integratics, economics integratics, management integratics, social integratics, etc.

These special integration theories construct a large group of disciplines. General integration theory is the integration of all these special integratics. In the forthcoming Part III, we will choose several fields and discuss the special integratics in these fields respectively.

In recent years, integration phenomena, as a kind of complex phenomena, have attracted some scholars. They have discussed some integration phenomena in some fields and proposed their viewpoints. Some of their work will be introduced as follows. (Due to the lack of materials, some work might be missing).

Liu Xiaoqiang (1997) proposed that the object of his study includes two aspects: one is all kinds of integration, such as information integration, technology integration, system integration, functional integration, process integration, environment integration, human and organization integration, etc., and the other is the interactions between each kind of integration. He believed that the research contents of integration include the classification, forms, conditions of production, mechanisms of formation, principles, rules and methods, and the relations of each kind of integration. He also proposed several research directions, such as human integration, comprehensive integration, analysis of complex giant system, modeling and simulation, and basic research of integration.

Hai Feng, Li Biqiang, and Feng Yanfei (2001) analyzed the nature, basic issues and categories of integrations based on system theory and extensive analysis of integration phenomena in economic and social organizations. They believed that the goal of integration is to investigate the theories of integration conditions, mechanisms and rules, and they proposed the basic issues and categories of integration unit, mode, interface, condition, environment, etc.

In addition, some scholars have studied some integration phenomena in specific fields, and proposed corresponding theories, such as Niu Shisheng's (1997) research on life integration, Hai Feng and Li Biqiang's (1999) study of management integration, Hu Qiyong's (2002) research on culture integration, Bao Hanfei's (2003) study of biological and medical knowledge integration, Li Biqiang and Hu Hao's (2004) research on enterprise property integration, Yu Hongyang, Li Haiyin and Lü Xin's (2005) research on network organization integration, and Chen Jiena and Wu Qiuming's (2007) research on industrial cluster integration, etc.

The aforementioned works have discussed many kinds of integration phenomena from different perspectives. Although their works do not provide the system of general integration theory, their survey, analysis, and discussion have provided materials about integration, contributing to further research on general integration theory.

6.3 Relations and Differences Between General Integration Theory, System Theory and Other Theories

In the 20th century, von Bertalanffy's general system theory, Wiener's cybernetics, Shannon and Weaver's informatics, Piaget's structuralism, and Qian Xuesen's theory of open and complex giant system were published in succession. These theories have deeply studied system, control, information, structure, open and complex giant system, illustrating many important concepts such as system, control, information, structuralism, comprehensive integration, etc. These theories have influenced the 20th century's science and technology, and enlightened the study on general integration theory.

This section will explain the relations and differences between general integration theory and these theories.

Von Bertalanffy has conducted long-time research on general system theory. His representative work is *General System Theory — Foundation, Development and Applications* (von Bertalanffy, 1987).

General system theory defines system as a set of related elements. It focuses on the models and principles that can be applied to general systems. It holds that the main feature of a system is integrity, which is larger than the sum of parts.

General integration theory has connections with general system theory. In general integration theory, there are concepts of integrated system and integrated unity. However, they concern different issues. General integration theory focuses on integration phenomena, especially integrative actions and integration processes, without concerning the features of a system. It emphasizes that integration is a process and it studies integration phenomena from the perspective of dynamic process.

Wiener et al. have systematically studied cybernetics. His

representative work is *Cybernetics—Or the Control and Communication in the Animal and the Machine* (Wiener, 2007). The cybernetics focuses on information exchange and communication process between system and environment or in an internal system, especially the feature of automatic adjustment. The main concepts proposed by cybernetics are control and feedback. He believed that the function control of system towards environment is realized by feedback. The mechanism of feedback is the basis of purposeful behaviors of animals and machines.

There are some connections between general integration theory and cybernetics. In general integration theory, there are concepts of regulation and control in the integration process. However, the integration process is very complex. The features of regulation and control are only a part of integration process. In addition, the two theories consider different problems. General integration theory discusses integration phenomena, not only involving regulation and control but also other mechanisms in the integration process, such as complementarity, synchronization, optimization, emergence and so on.

Shannon and Weaver have deeply studied informatics. Their representative work is *The Mathematical Theory of Communication* (Shannon & Weaver, 1949).

The main concept of informatics is information, defined by negative entropy in thermodynamics. Informatics discusses the reception, transmission, coding and exploitation of information.

General integration theory is connected with informatics. It discusses information processing in the integration process and also studies information integration. However, information processing and information integration are only one type of process among all kinds of integration phenomena of general integration theory. General integration theory discusses not only information integration, but also structural integration, functional integration, psychological integration, social integration, etc.

Piaget discussed the structuralism of scientific understanding. His representative work *Structuralism* (Piaget, 2006) is a part of his genetic epistemology.

Structuralism can be dated back to the concept of synchronic system in linguistics by de Saussure (de Saussure, 1980) and the concept of perceptual field of gestaltism in psychology (Hothersall, 1984).

Structuralism has investigated the structure of scientific understanding in many different disciplines, such as mathematical structure, physical structure, biological structure, psychological structure, linguistic structure, the application of structure in social research, structuralism and philosophy. Structuralism holds that structure has three features: integrity, convertibility, and self-regulation.

General integration theory is related with structuralism. General integration theory discusses the problem of structure, especially structural integration and theory integration in integration phenomena. However, the general integration theory involves a wide range of investigation, including structural integration, theory integration and scientific understanding.

Qian Xuesen has done pioneering work in open and complex giant system. His representative works are *About Scientific Understanding* (Qian, 1986), and *Create a Systematics* (Qian, 2007).

He studied artificial intelligence, putting forward the metasynthesis from qualitative method to quantitative method of open and complex giant system. This method combines experts, data and information, and computer hardware and software. These three parts construct an integration system.

Based on that, he further proposed the hall for workshop of

metasynthetic engineering, which is to adopt human-machine combination, synthesizing millions of people's intelligence, including thoughts, fruits of human thinking, disciplinary knowledge, experience, and all kinds of intelligence information to solve complex problems. His ideas provide profound guidance for future artificial intelligence system of human-machine integration.

General integration theory and the open and complex giant system theory have connections, with both involving the concept of integration. However, they discuss different problems. The latter considers artificial intelligence, especially the people-friendly human-machine combination from the perspective of engineering and technology, and the metasynthesis from qualitative to quantitative method, while general integration theory is based on learning from the brain, focusing on integration phenomena, especially the general features, rules and applications of integrative actions and integration processes.

To sum up, general integration theory has many connections with the aforementioned theories, but they are different. Their differentiation is not about opposite opinions, but research objects, contents, concepts, methods and application.

For research objects, the former theories have a general system, control process, information principle, knowledge structure, open and complex giant system as research objects, while general integration theory takes integration phenomena in nature, technology and human society as its research objects.

For research contents, the former theories study the rules of system, control, information, structure, metasynthesis, while general integration theory conducts researches on the general features and rules of all kinds of integration phenomena in nature, technology and human society, and also the integrative actions and integration processes of all kinds of structural integration, functional integration, information integration, psychological integration, knowledge integration, environment integration, etc.

As for the concepts, these theories respectively discuss the concepts related to system, control, information, structure, artificial intelligence, while general integration theory discusses the concepts related with integration. Although general integration theory also involves these concepts, it mainly focuses on integration, coordination, optimization, complementarity and synchronization.

As for the research methods, they are different. General integration theory begins with learning from the brain and further studies integration phenomena of other fields based on large amounts of experimental facts. However, the aforementioned theories do not study the brain, and thus do not involve the structure, function, information processing and advanced activities of the brain.

As for the range of application, general integration theory can be applied to different fields, to discuss integration phenomena in many specific sub-fields and thus develop a large group of sub-disciplines, such as bio-integratics, med-integratics, psycho-integratics, engineering integratics, technology integratics, education integratics, social integratics, etc. This is different from the other aforementioned theories.

All in all, general integration theory and the aforementioned theories are connected but different. They are complementary to each other. We hope that general integration theory and its further studies can be based on former theories and meanwhile complement and enrich them.

Part III

Applications of General Integration Theory

To apply the theories of general integration theory to different fields, we should study integration phenomena and their features and rules in different fields, for example, bio-integratics, psycho-integratics, technology integratics, engineering integratcs, education integratics, social integratics and so on. In the following chapters, we will discuss bio-integratics, psycho-integratics, knowledge integratics, engineering integratics and education integratics. They are examples of the applications of general integration theory.

Another large research field is to apply general integration theory to social systems, studying all kinds of integration phenomena in social systems. We call it social integratics. It includes the studies of integration processes of all kinds of social phenomena, such as management integratics which studies the integration phenomena, rules and applications in the management process, team integratics which studies the integration phenomena, rules and applications in the team organization and so on.

This kind of research has practical meaning. In Chapter 6 we have introduced current works about the management process and team organization. It is worthwhile to undertake systematic and deep research on the basis of these works (Gintis & Bowles, 2005).

Chapter 7 Bio-integratics

On earth, there exist all kinds of creatures. They are undergoing colorful life activities. In Chapter 4 we have already mentioned that the biological world has hierarchical structure, from biomolecule to the whole biological world. In life activities at different levels, there are integration phenomena with all kinds of features.

To apply the general integration theory to all kinds of integration phenomena in the biological world will help us understand the features of creatures and their life activities at different levels. Thus, we want to build a discipline to study the integration phenomena in the biological world, their rules and applications, and we call it bio-integratics.

Each kind of creature is the product of evolution. The integration phenomena of creatures at different levels have their own features. Thus the range of research on bio-integratics is very broad. This chapter only chooses three issues respectively: the bio-integration and biological evolution, the integration in living cells, and the integration in human bodies.

7.1 Bio-integratics and Biological Evolution

There are many levels in the biological world. Biological macromolecules, such as nucleic acid and protein, are basic materials for constructing a living body. There are many other biological macromolecules, such as sugar and membranes. The cell, which is constructed by biological macromolecules, ions, hydrone, is the unit of all kinds of creatures. Multi-cellular creatures, such as animals and plants, have complex structures. The biotic community and environment construct the ecosystem.

Creatures and their activities of different levels have different scales of time and space. For example, the spatial scale of biological macromolecules is about 10^{-8} meters, with a time scale of about 10^{-9} seconds; the spatial scale of cell is about 10^{-6} meters, and the time scale is about 10^{-2} seconds; the spatial scale of multi-cellular creatures is about 10^{-1} meters, and the time scale is about 10^{-1} seconds; the spatial scale of ecosystem is about 10^{2} meters, and the time scale is about 10^{2} seconds.

Chapter 4 has mentioned the integration phenomena in the biological world. Creatures of all different levels are all the products of integration. For instance, the biological macromolecule is integrated by micromolecules; cell is integrated by biological macromolecules; multi-cellular creature is integrated by cells; ecosystem is integrated by creatures and environment (Chang & Ge, 2005).

The most interesting research area in biological integration phenomena are the integrative actions and integration processes of creatures at different levels. In the process of biological integration, the interactions of creatures are very important. Through the interactions, different integration components are integrated into the creatures. There are several interactions of different features, including the interaction between the internal components of creatures, the interaction between different creatures, the interaction between creatures and environment. These interactions are reciprocal. The interaction between creatures and environment is an example. On the one hand, the environment influences the creatures and changes them; on the other hand, the creatures place impact on the environment and change it, too.

Each creature has its internal life activities, such as metabolism and development. Every creature lives in a certain environment with other creatures. There are always exchanges of materials, energy and information between creatures and environment. One creature is always communicating with other creatures. For any one creature, there are internal integration processes and also external integration processes. Inside the creature, different components are integrated into a unified organism, which is realized by the interaction of the internal elements of the creature. The creatures and the environment are integrated into a unity, which is realized by the interaction between creatures and between the creatures and the environment.

There are all kinds of creatures in the biological world. Darwin's theory of origin of species uncovers the rules of biological evolution and natural selection, pointing out that the current creatures have all experienced long-term evolution. Modern people have evolved from advanced primates. From ancient apes to modern people, it has lasted about 10 million years (Eccles, 2004).

The neural system of modern people is also the product of evolution. It has intricate sensation, perception, delicate motion and control, and all kinds of advanced functions. They are all the results of long-term evolution and natural selection.

Eccles's (2004) book *Evolution of the Brain* collected lots of scientific proof about the evolution of ancient ape to human beings in the fields of archeology, neuroanatomy and brain physiology. It describes the main features of evolution, studies the history of brain evolution and explored the origin of consciousness of higher animals.

For those who lived in the environment during ancient times, upright and biped walking are very important in the process of brain evolution, because when people walk straight on two feet, they need the delicate control of all kinds of muscles by the brain cortex. Human labor has a decisive role in the development of the brain. For example, to make use of wood and stone requires the exquisite movement of the hands, which promotes the evolution of the motor area of the brain cortex and its neural connections. To sum up, the structure and the function of the brain are always developing in the process of evolution. Therefore, the human brain is the result of long-term evolution.

There are many integration phenomena in biological evolution. These integration phenomena are long-lasting. The integration goals and steps are not predetermined but realized step by step during the interactions of creatures and environment. The integration process of biological evolution is the process of patching.

7.2 Integration in a Living Cell

A cell is the product of evolution. It includes the prokaryotic cell and the eukaryotic cell. The prokaryotic cell has no nucleus while the eukaryotic cell does. Here we only discuss the eukaryotic cell.

The living cell is always undergoing all kinds of life activities. The integration in a living cell refers to the integration phenomena in the life activities of a living cell. The living cell has complex structures and functions, which are integrated by all kinds of cellular materials, namely nucleus system, cytoplasm system, membrane system and cytoskeleton system. The materials in cells are mainly composed of genetic materials, cytoplasm materials, membrane materials, cytoskeleton materials, etc.

The nucleus system is the genetic materials system with the main function of realizing heredity. The genetic material is the DNA molecules carrying genetic information, which concentrates in the nucleus. The nucleus has a membrane, which separates nucleus from cytoplasm, providing space for the storage and copy of genetic materials. Through the membrane of nucleus, there is constant transfer of materials, energy and information inside and outside the nucleus. The nucleus is integrated by all kinds of components. During the period of mitosis, the nucleus undergoes reintegration.

The cytoplasm system has many materials, such as biological molecules, ions and hydrones. They are integrated into many kinds of organelles and the whole cell. The main function of a cytoplasm system is to realize all kinds of life activities of a cell. In this system, there is material transportation, energy metabolism, and information exchange. During the period of mitosis, the cytoplasm undergoes reintegration.

The membrane system includes plasma membrane and nuclear membrane, and also many subcellular structures with membrane. The main function of this system is to provide the space of separating materials and the interface of communication between the inside and outside materials. For example, plasma membrane, namely cell membrane, forms the interface between the inside and outside of the cell, providing the living space of the internal part of the cell and promoting the communication between the inside and outside of the cell. Plasma membrane has the structure of lipid bilayer. There are proteins on the membrane to form the receptor, ion pathway and water pathway. The plasma membrane is mainly integrated by these parts.

The meshwork supported fluid film model of cell membrane puts forward that the membrane includes lipid bilayer and membrane skeleton. The lipid bilayer has two layers with the upper layer embedded with proteins while the membrane skeleton under the lower layer is constructed by skeleton proteins to support the fluid membrane and provide network for transporting membrane materials (Tang, 1998).

The main components of cell skeleton system include microfilaments, microtubules and intermediate filaments. They are integrated into the fibril structure. The main function of this system is to support the cell and to ensure the internal material transportation. For example, the particles inside the cell, acted on by motor proteins, can transfer along the cell skeleton.

In a series of experiments, the cells of water mold were used to study the structure and function of skeleton structure inside the cell. Exploiting the method of electron microscope and optical microscopic video, the fibril structure and its features are measured, including the phenomena of the moving particles in the cell along the fibril. The results show that inside the living cell, there is a delicate transportation system composed of cell skeleton (Yan, Tang, & Liu, 1994).

These four structure and function systems of the cell are all integrated by biological macromolecules. Inside the cell, these systems are not independent, instead, they have close connections. The whole living cell is integrated by these four systems. The coordination of these four systems ensures the normal life activities of a living cell.

In the process of development and division of an eukaryotic cell, there are many special integration phenomena. The cell has a cycle of division, including interphase and division phase. The cell grows during interphase. The material integration during interphase is one kind of cell integration. In the division phase, the cell produces daughter cells. The integration phenomenon in the process of producing daughter cells is another form of cell integration. This process constructs daughter cells based on internal copy and material division.

The mitosis of eukaryotic cell includes prophase, prometaphase, metaphase, anaphase A, and anaphase B. During prophase, the cell constructs chromosomes and a spindle with two poles. During prometaphase, the chromosomes move towards the equatorial plane of the spindle. During metaphase, all the chromosomes line up on the equatorial plane, then vibrate up and down, and then split into sub-chromosomes. During anaphase A & B, the sub-chromosomes move to the two poles of the spindle, meanwhile the cell splits into two daughter cells.

During mitosis, the activity of chromosome is influenced by the acting force, which is also an important part of the integration process. We once studied features of the acting force on the chromosomes. According to the experimental data of the movement of sub-chromosomes towards the two poles of spindle, it can be shown that the force making the sub-chromosome movement is not in direct proportion to the length of connecting fibers; instead, it keeps constant (Tang, 1992).

In the process of mitosis, the cell integration is undergoing. During anaphase B, when sub-chromosomes arrive at the two poles of the spindle, the integration of the daughter cells is complete. The daughter cell is integrated by the sub-chromosomes, and the distributed cytoplasm and plasma membrane. Meanwhile, there is reconstruction of the nucleus of daughter cells, which is also an important part of the integration of daughter cells. The result of cell division is to form two relatively independent daughter cells.

There are many experimental researches on the process of cell integration. For example, in an experiment, the egg of xenopus was used to specifically study the reconstruction of nucleus of the cell (Yang et al., 2003). The dynamic image of the reconstruction of nucleus has been observed by the scanning atomic force microscope. This study has discovered that the reconstruction of cell nucleus is one important step in the process of cell integration.

In an experiment, the pollen tube was used to study the phenomenon of apical development of the living cell (Tang et al., 1992). It was observed that in the process of apical development of the pollen tube, the materials inside the pollen tube are constantly transported from cell body to the tip to make the tip grow. This kind of development is not successive but abrupt. This study shows that the apical development of the pollen tube is one kind of cell integration.

7.3 Integration in Human Body

Human body is a complex biological system, including a circulatory system, a respiratory system, a digestive system, a neural system, a skeletal system, and a genital system. The circulatory system has the function of blood circulation; the respiratory system has the function of inhaling oxygen, and exhaling carbon dioxide; the digestive system has the function of digesting food and absorbing nutrition; the neural system has the function of controlling and coordinating; the skeletal system has the function of supporting the body and ensuring the movement; the genital system has the function of breeding the descendent. Besides, there is an endocrine system and an immune system. These systems coordinate to form the unified human body. In the human body, there is internal communication between materials, energy and information, and meanwhile, there is communication between human body and external environment.

From the perspective of bio-integratics, a human body is integrated by these sub-systems. There is not only structural integration, but also information integration. Human body is considered as a whole. Sherrington (1906) discussed the mechanism of integrating each part of the human body into a whole in his book *The Integrative Action of the Nervous System*. He believed that each part of the human body is integrated by neural activities, blood circulation, endocrine process and so on, among which the integrative action of neurons is the most important.

According to Chinese traditional medicine, the human body has a system of main and collateral channels. Our paper *A Hypothesis About the Main and Collateral Channels of Human Body* (Tang, Shen, & He, 2008) has discussed this problem. From the viewpoint of human body integration, the main and collateral channels of the human body might be one way of internal integration of the human body. Some explanations are cited as follows:

> The long-term practice of Chinese traditional medicine has proved the medical effect of acupuncture. Large amounts of clinical observations and experimental data have shown that acupuncture can adjust the function of systems and body organs on many aspects, with many steps, at multiple levels and in many ways.
>
> The acupuncture cannot function without the accurate acupoints. Traditional Chinese medicine believes that an acupoint is the point where the main and collateral channels go in and out of the surface

of the body. They are not independent points, but special parts which have intricate relations with internal organs. They are also the place of spotting the illness. Each acupoint has a specific relation with internal organs. The channels provide these relations for us.

The acupoint is connected with internal organs. The dredging function is bidirectional. To have acupuncture or moxibustion on the surface of human body can treat the illness on the main and collateral channels. The physiological status of internal organs and their pathological change can also be reflected on the acupoints through main and collateral channels. In the pathological status, some acupoints often undergo specific changes. For example, patients with gastroenteropathy usually have obvious pains at acupoints Zusanli and Di chi. Patients with lung disease usually have obvious pain and subcutaneous nodule at lung stream points and acupoint Zhongfu. The reflection of organ disease at the corresponding lung stream points is mainly completed through main and collateral channels. The main representations are pressing pain, bitterness, induration and so on. These facts are helpful for sickness diagnosis and choice of effective acupoints.

The way of acupuncture can be used to treat the illness of relevant internal organs. The illness of internal organs also shows abnormal change at the corresponding acupoints. These organs which have corresponding acupoints are called targets. This kind of correspondence is based on main and collateral channels.

The knowledge about main and collateral channels is constantly developed in practice, which has become one of the most important contents of Chinese traditional medicine and also the basis of acupuncture. Main and collateral channels have the functions of connecting, reacting and adjusting. It is the most important part of human body. It combines with internal organs to form the human body. It is widely distributed, connecting human body into a unified organic whole through regular circulation and complex network. The system of main and collateral channels includes main channel and collateral channel. Main channel is the trunk, buried deeply and connected with the whole body. Collateral channel is the branch, existing on the surface, all over the body. The system of main and collateral channels closely links the organs all over the body, playing an important part on physiological, pathological aspects and prevention of disease.

Scientists have observed and studied the main and collateral channels and its nature in many aspects, but have not made conclusions about the material basis of main and collateral channels.

Human body is a complex system integrated by many systems, among which the neural system, the endocrine system and the immune system compose the overall system that has the function of regulation. In recent years, there are many researches about the molecular mechanism of this overall system. These studies show that these three systems form a whole network.

The neural-endocrine-immune system is an important part of the union of mind and body. From the perspective of physiology and psychology, the interaction between mind and body is realized by neural-endocrine-immune system. The neural system includes the central nervous system and peripheral nervous system. In the endocrine system, the endocrine gland releases the hormone, influencing the activities of effectors. Hormone is a kind of chemical materials secreted by the endocrine gland, such as epinephrine, noradrenaline, and cortisol. The immune system produces antibody to fight with pathogeny from outside. The antibody can recognize and resist the foreign matters inside the body. For example, the antibody in blood and body fluid can kill and restrain the bacteria.

In the neural-endocrine-immune system, there is transfer of neural signals and chemical materials, including all kinds of hormones and neurotransmitters. These chemical materials will combine with receptors and have an impact on the function of immune system. In the books written by McCann et al. (1998) and Melmed (2001), there are detailed illustrations about the physiological processes and molecular interactions at the interface of the neural system, the endocrine system and immune system.

The neural signals can directly control the activities of effectors, and also regulate and control the effectors through endocrine gland. For example, when someone is excited, the signal of his brain will trigger the reaction of automatic nervous system to regulate the activities of endocrine system. The adrenal gland secretes cortical hormone, transported to each part of body through the blood. The increase of cortical hormone will restrain the activities of immune system and influence the ability of fighting with the sickness. The change of immune system in turn will have an impact on the neural system, which has been studied by Maier and Watkins (1998).

The delivery of endocrine hormone is accomplished through blood flow. Because the transmitting speed of chemical materials is slower than that of neural signals, the lasting time of chemical materials is longer than that of neural signals.

Psychoneuroimmunology shows that in the neural-endocrine-immune system, if any one system is in disorder, the other two systems will be influenced. For example, the disorder of neural system will result in the imbalance of endocrine system and the decline of the functions of immune system. In the immune system, if the process of producing antibody is obstructed, the immune function of human body will be in disorder.

Usually, the neural-endocrine-immune system is all over the body and diffusely distributed. Although this view can explain many physiological phenomena, it can hardly directly explain the fact that acupuncture on certain acupoints of a human body will be effective. We want to improve this view and propose the hypothesis of neural-endocrine-immune network with sensitive nodes and functional connectivity.

In nature, there are lots of complex systems that can be described through all kinds of networks. A classical network is composed of many nodes and some edges connecting the nodes. The nodes represent different individuals in a real system, while the edges represent the relations of those individuals. Usually, if there is a specific relation between two nodes, the two nodes are connected. For example, a neural system is a network that is integrated by lots of neural cells through nerve fibers. A computer network is the network that is integrated by computers, optical cables, twisted pairs, and coaxial-cables. Similarly, there is a power network, a social relation network, a food chain network, and so on. The neural-endocrine-immune system is also a kind of complex network.

The hypothesis of the neural-endocrine-immune system with sensitive nodes and functional connectivity has the main points as follows (Tang, Shen, & He, 2008):

First of all, the neural-endocrine-immune system is a complex network inside the body. This network is distributed all over the body, and has a series of sensitive nodes and the functional connectivity between the sensitive nodes. The functional connectivity results in temporal correlations between the sensitive nodes which have spatial distance. It differentiates from the structural connection, but is based on it, reflecting the similar responses of different structures towards neural and physiological events.

Secondly, the functional connectivity is realized by the transfer of neural signals and chemical materials.

Thirdly, the physical stimulus (acupuncture, moxibustion, electrical stimulation) imposed on the sensitive nodes of this complex network can regulate the network. Different sensitive nodes relate with the relevant targets to treat corresponding illness. A series of acupoints are the sensitive nodes of complex network.

We have proposed the hypothesis based on the view of integration of human body, combining the main and collateral channels and the neural-endocrine-immune system with sensitive nodes and functional connectivity. If the main and collateral channels in Chinese traditional medicine belong to this system, the sensitive nodes in the network are equivalent to a series of acupoints in the human body, the functional connectivity between sensitive nodes is equivalent to the main and collateral channels of the acupoints, then the concept of human's main and collateral channels can be consistent with the neural-endocrin-immune system. Of course, this complex network includes not only the sensitive nodes and functional connectivity, but also collateral pathways over the body.

From the perspective of physiological functions, if human main and collateral channels are equivalent to the part of sensitive nodes and functional connectivity in this complex system, the functions of main and collateral channels can be related with the physiological functions of the neural-endocrine-immune system. This complex network is not only distributed all over the body, but can also regulate the body through the sensitive nodes and functional connectivity.

The complex networks have spatial structure, and the interactions in the network also have a temporal dimension. According to the former hypothesis, the treatment effect of acupuncture also has temporal and spatial features. The spatial feature of the treatment effect of acupuncture is that in the complex network, there are definite sensitive nodes with corresponding targets, which can treat specific illness. The temporal feature of acupuncture is that the treatment includes fast component and slow component. The fast component is the neural signals that activate the network when people are acupunctured. The neural signals directly control the target. The neural signals are transmitted fast and the lasting time is short, thus the regulation is temporary. The slow component of acupuncture is the effect of endocrine and immune signals. The delivery speed of endocrine hormone is slower than neural signals, but lasts longer, thus might exert lasting treatment effects.

Chapter 8 Psycho-integratics

There are all kinds of integration phenomena in the colorful mental activities. It is very interesting to apply the viewpoints of general integration theory to study the features of all kinds of integration phenomena in psychological activities. We want to build a new discipline of psycho-integratics, which studies the integration phenomena of psychological activities, rules and applications.

There are many integration phenomena in psychological phenomena. Chapter 2 has introduced all kinds of binding phenomena in psychological phenomena. These are classic examples of psychological integration. Integration in the thinking process is also a kind of representation of psychological integration. This chapter will discuss several aspects of integration phenomena of psychological activities, i.e. mental interactions and psycho-integration, integration for consciousness, integration for cognition, mental integration, and integration of mind and behavior.

8.1 Theory of Mental Interactions and Psycho-integration

The mental interactions are very important to the integration phenomena of psychological activities. The integration process is realized by all kinds of mental interactions. Thus we need to study the features of all kinds of mental interactions in psychological phenomena and how the different psychological integrations are realized by mental interactions.

The psychological phenomena include all kinds of mental activities and behaviors. The individual psychological phenomena not only include internal mental activities, but also relate to many elements of the brain, body, natural environment and social environment. The brain, body, natural environment and social environment, are not independent. They interact with one another and form an integrated whole.

Enlightened by the studies of physical interactions and their unification, we have proposed the concept of mental interactions and their unification in psychological phenomena. We have discussed all kinds of problems about mental interactions in psychological phenomena in the book *Unified Framework of Psychology and Cognitive Science* (Tang, 2007). We call the interactions in psychological phenomena as mental interactions. There are several different kinds of mental interactions in the colorful psychological phenomena. A mental interaction is the interaction between psychological activities and a certain element, including the effect of this element on psychological activities and the effect of psychological activities on the element. The relation between individual psychological activities and social environment is an example. The social environment has an effect on the individual psychological activities and in turn the individual psychological activities have an effect on social environment. There is interaction between them.

The colorful psychological activities are undergoing in the human brain. The brain and body are in the ever-changing environment. In the ever-changing environment, many things act on the individual brain and body; meanwhile the brain and body react on these things. Thus, when we study the individual psychological activities, it is necessary to consider the relations between the internal elements of the mental system, and also the relations between the individual psychological system and all kinds of external elements.

When we investigate the relations between the internal elements of individual mental system in psychological activities, we see several mental interactions of different features, including the interactions between each kind of element in psychological activities and the interaction between certain psychological activities and the brain.

Psychological activities include sensation, perception, learning, memory, attention and thinking. Psychology has definitions on these processes (Sdorow, 1995; Peng, 2001). For example, learning is a change of behaviors and behavior potentials produced by repetitive experience in a specific situation. Memory is the process of accumulating and storing individual experience in the brain. Thinking is the general and indirect understanding about objects with the help of language, representation and behavior. Inference is to conclude general rules from concrete things.

It has been mentioned before that the psychological activities have components of arousal-attention, cognition, emotion, will, and so on. These components are not independent, but closely connected, and constantly interacted. Each kind of component in the psychological activities interplays with one another. For example, the emotion acts on cognition and in turn the cognition reacts on emotion, which is seen as the interaction between emotion and cognition.

Mental integration is the integration of all kinds of psychological components, which is realized by the interactions among them.

As far as the relation between psychological activities and the brain is concerned, it should be said that the brain is the basis of psychological activities, while psychological activities play the function of the brain. Each kind of component of psychological activities is the brain function instead of certain entities. Thus the interactions between the psychological activities are not the interactions between the entities. However, these psychological components are all based on the brain. The brain mechanism of interactions between the components of psychological activities is the interactions among the functional systems inside the brain. The psychological activities have interactions with the brain, including the effect of psychological activities on the brain and the effect of the brain on psychological activities. These interactions are called mind-brain interaction. The mind-brain system is the integration of mentality and brain, which is realized by the interactions between psychological components and those between mentality and the brain.

The relation between mind and brain is the relation between psychological activities and the brain. The mind-brain interaction is an important content of mind-brain relation. The psychological activities cannot happen without the brain. The interaction between the brain and mind is the interaction between psychological activities and the brain, thus the nature of mind-brain interactions is different from the interactions between psychological components.

We further consider the relation between mind-brain system and each kind of external elements in individual psychological activities. We can see the interactions between psychological activities and external elements at all levels. They are psychological interactions of different properties. These interactions are mutual.

As to the relationship between psychological activities and body, each kind of physiological signals related to the psychological activities is delivered in the neural-endocrine-immune system, integrating psychological activities and body movement. For example, the stimulus of external environment generates neural signals on the reception organs, being transmitted to the brain through neural system, and resulting in all kinds of feelings and perceptions. The neural signals are delivered from the brain to each part of the body to control the movement of these organs. The interactions between psychological activities and body include the effect of psychological activities on the body and the effect of the body on psychological activities. The mind-brain-body system is formed by the integration of mind, brain and body through interactions between psychological components, between brain and mind, and between body and mind.

The mind-body relation is the relationship between psychological activities and body. The brain is an organ in the body and a part of the body. The psychological activities cannot proceed without the brain and the brain cannot live without the body. The interaction between mind and body is an important content of mind-body relation. In this interaction, the mind is not an entity but the function of the brain.

Individuals always live with the external environment. There are interactions between individual psychological activities and natural environment. On one hand, the psychological activities produce action through the brain and body and act on the natural environment; on the other hand, the natural environment constantly stimulates the individuals, acting on psychological activities through body and the brain. This is an interaction between the psychological activities and the natural environment. The mind-brain-body-environment system is integrated by the interactions between psychological components, between the brain and mind, between mind and body, and between mind and natural things.

The mind-matter relation is the relation between psychological activities and matters in the natural environment where the individuals live. The mind-matter interaction is an important content of mind-matter relation. The mind-matter interaction is different from the mind-body interaction. In the mind-matter relation and mind-matter interaction, the mind is not a kind of entity but a function of the brain.

Individuals live in a social environment, so there is an interaction between psychological activities and social environment. On one hand, psychological activities produce movement through the brain and body and act on the social environment through brain and body; on the other hand, the social environment constantly stimulates individuals and acts on psychological activities through body and the brain. These interactions are called the mind-society interaction. The mind-brain-body-society environment system is integrated by the interaction between psychological components, mind-brain interaction, mind-body interaction, mind-matter interaction, and mind-society interaction.

The individual psychological activities have close relations with the social environment. The individual mind-society interaction is an important part of the relation between individual mentality and social environment. The mind-society interaction is different from other psychological interactions. In this interaction, the psychological activities are not a kind of entity, but the function of individual brain.

As indicated above, there are all kinds of relations of psychological activities, such as the relation of each kind of psychological activities, the relation between psychological activities and brain, the relation between psychological activities and body, the relation between psychological activities and natural matters, and the relation between psychological activities and social environment. In the unity of mind, brain, body, natural environment and social environment, there are networks of many levels and aspects.

To sum up, there are five kinds of psychological interactions in psychological phenomena. They are the interactions between psychological components, the mind-brain interaction, the mind-body interaction, the mind-matter interaction, and the mind-society interaction. The psychological interactions in psychological phenomena are very complex. In a psychological phenomenon, usually there are many kinds of psychological interactions instead of only one. The integration process in the unification of mind, brain, body, natural environment, and social environment is realized by the aforementioned five kinds of psychological interactions.

The book *On Intelligence — Integration of Mental and Behavioral Abilities* (Tang, 2010) pointed out that these psychological interactions have different spatial and temporal ranges, different ways and different results of interactions. These psychological interactions undergo at different levels of the unity of mind, brain, body, environment, and society. However, they are all based on the activities of mind-brain system, thus they can be unified on the basis of mind-brain unity.

One feature of psychological interactions is the reciprocity. For instance, there are two aspects of the mind-brain interaction. On the one hand, the electric activities and chemical reactions in the neural system provide us with the biological basis of psychological activities. They determine the psychological activities and each kind of psychological activities has a corresponding brain mechanism. On the other hand, the psychological activities are accompanied by electric activities and chemical reactions, which can shape the neural system inside the brain. Thus the interaction between psychological activities and the brain is reciprocal.

Another feature of psychological interactions is the dynamic course. The individual brain and mentality develop by the joint action of all kinds of psychological interactions. In the brain, there are constant psychological activities. The action of the brain on psychological activities and the reaction of psychological activities on the brain make the mind-brain system coordinate. As the brain has plasticity, the individual mind and the neural network in the brain can constantly develop in this dynamic state.

8.2 Integration for Consciousness

Consciousness is a complex mental phenomenon. The integration of consciousness is a kind of complex psychological integration phenomenon. We should study the integration of consciousness from many different aspects. The state of awareness and the experience of awareness are two important aspects of consciousness. This section will investigate the integration of consciousness from the integration of the state of consciousness and the experience of awareness.

Searle (2000) said that consciousness has a series of properties, such as qualitative, subjective, unified and fluid when he discussed the features of consciousness. The qualitative feature refers to the specific qualitative experience when people are conscious; the subjective feature refers to the individual subjective experience when people are conscious; the unified feature refers to the individual experience of consciousness which is integral; the fluid feature refers to the individual experience of consciousness which is constantly updated.

The subjective, unified and fluid features are all closely related with the integration of consciousness. The subjective feature of consciousness shows that the integration is inside the individual mind-brain system. The unified feature of consciousness shows that the overall experience is the result of integration of consciousness. And the fluid feature of consciousness shows that consciousness is the process of integration.

To consider the conscious activities from the perspective of the state of awareness, the whole state of awareness is integrated by consciousness, unconsciousness and latent-consciousness. Mental activities include conscious and unconscious activities and the mental structure also has the components of latent-consciousness (Tang, 2008a).

Based on whether the information processing accesses individual consciousness or not, the information processing might be conscious or unconscious. The former kind of information processing takes part in conscious cognitive activities and is explicit. Information is processed in abundance in which the activation level of relevant brain regions is lower than the threshold of consciousness, thus cannot access individual consciousness. Although the unconscious information processing cannot be perceived by individuals, they also participate in cognitive activities, which belong to implicit information processing (Tang, 2004).

Some information processing inside the brain cannot be perceived by individuals because the relevant neural activities have no relation with perception systems, for example, the information storage process, the information transfer process and the steps of information processes. They are called non-conscious nervous activities. The specific information inside the brain is stored by the corresponding brain regions. When the brain regions are not activated, the information is not retrieved and processed. The state of information which is stored while not retrieved and processed is called latent-consciousness (Tang, 2008a).

The previous mentioned states, i.e. consciousness, unconsciousness and latent-consciousness, are all states of awareness. Our definitions of unconsciousness and latent-consciousness are different from other researchers. Some researchers regard the unconsciousness mentioned here as latent-consciousness, and call the latent-consciousness mentioned here as the trace of memory.

In the book *On Consciousness—Natural Science Research on Consciousness* (Tang, 2004), we regarded the different states of consciousness as the different energy states of consciousness. To be more specific, the latent-consciousness is regarded as the ground state of awareness, the unconsciousness as the excited state of low activation with the activation level lower than the threshold value, and the consciousness as the excited state of high activation with the activation level higher than the threshold value.

The integral state of consciousness is integrated by the ground state, the excited state of low activation, and the excited state of high activation. Under certain conditions, there will be transitions between different energy states. For example, the conscious retrieval of information is achieved by the transition from the ground state to the excited state of high activation, the change from the unconscious information processing into conscious information processing is achieved by the transition from the excited state of low activation to high activation, and so on. In the theory of global workspace of consciousness proposed by Baars (1998), the theory of neural global workspace of consciousness proposed by Dehaene (2001) and the theory of expanded global workspace of consciousness proposed by us (Song & Tang, 2008), the process of unconscious activities entering global workspace and transferring into conscious activities has been discussed.

The consciousness as complex psychological phenomena has internal structures. The conscious experience has some basic elements, such as the arousal of consciousness, the content of consciousness, the intention of consciousness, and the emotion of consciousness. Lots of experimental facts show that awareness is connected with arousal of consciousness, that consciousness has contents, and that the process of consciousness is also accompanied by intention and emotion. These four elements of consciousness respectively reflect the important features of conscious experience. They all have psychological meaning. To see the conscious activities from the perspective of conscious experience, the whole conscious experience is integrated by the experience of arousal, content, intention and emotion. We have discussed these elements of consciousness in the book *On Consciousness—Natural Science Research on Consciousness* (Tang, 2004).

The arousal of consciousness is one element of the conscious experience. Consciousness has subjectivity, which is personal, and there are all kinds of subjective experience in the state of consciousness. There are different degrees of arousal, which reflect the intensity of conscious experience at some time. The individual can experience not only the state of his arousal, but also the degree of his arousal. A certain state of consciousness is the basis of conscious activities, and thus the arousal of consciousness is one of the elements of conscious experience.

The individual arousal of consciousness is related to the physiological state and is restrained by it. For example, the degree of the arousal is regularly changing with the time. It is different when the individual is awake or asleep, and it is also influenced by his physiological state.

The experience of consciousness is integral. The arousal of consciousness describes the overall state of individual conscious experience. It is fluid and changing. The degree of arousal sometimes is the feature of conscious experience at this very moment and reflects the average level of awareness during this time interval. Arousal is a necessary condition of awareness. The content of consciousness is another element of conscious experience. An individual can know not only whether he himself is aware or not, but also what he is aware of. What an individual becomes aware of is the concrete content of his subjective experience. Based on this experience, the individual can also know the meaning of his awareness.

The content of consciousness includes the events and knowledge that access to the consciousness. The individual conscious experience is colorful, for instance, experiencing some process or situation, or some concept or some thought. No matter what people become aware of, they have concrete content with psychological meanings. Because the conscious experience always includes some specific content, the content of consciousness is one element of conscious experience.

Consciousness is always changing, so is the content of consciousness. The content of events or knowledge accesses to the individual consciousness at some time, and at this moment, the individual experiences the content. The content of consciousness is information, and the alteration of content forms the information flow inside the brain. In the content of consciousness, there are all kinds of information being processed inside the brain, namely the content of psychological activities. Apart from the input information received by the brain and the corresponding subjective experience, the content of consciousness also includes understanding the meaning of information. In the content, there is also the output information given by the brain, namely the content of behaviors being regulated.

The individual intention and emotion both include information, and they also construct the broad content of consciousness. However, the content of consciousness discusses the information about certain events, things and knowledge and the understandings about them, which differentiates the content of consciousness from the intention and the emotion of consciousness. Thus, besides the content of consciousness, intention and emotion are another two elements of conscious experience.

Intention is another element of conscious experience. Intention has direction. The individual consciousness experience results in further intentions. Especially when the meaning of experienced content is understood, the intention will be clearer. Intention promotes the flow of content of consciousness, transmitting the content at this moment to that of the next moment. These intentions make individuals draw up goals, plans and promote further activities. The subjective experience of individual intentions has psychological meaning, so it is one element of conscious experience.

Emotion is another element of conscious experience. In psychology, the short-term emotion is called mood while the long-term emotion is called feeling. They both have psychological meanings. This kind of subjective experience is an important part of consciousness.

Each kind of element of consciousness includes many different types and features. For example, for the content of consciousness, there are two types: one is the content of events and things, such as the features of matters, the situation of events and their temporal and spatial features and the understanding of their meanings; another is the content of knowledge, such as all kinds of concepts and rules and the corresponding understandings.

Each kind of content of conscious experience includes many

specific features. On the aspect of the features for matters and the situation of events, in terms of different sensory channels, there are visual perceived content, auditory perceived content, gustatory perceived content, olfactory perceived content, tactile perceived content, etc. The content of consciousness includes perceptions with single feature and those with binding features. The latter is integrated by perceptions in all kinds of sensory channels. What is most discussed is the visual perceived content, also called the visual consciousness (Crick & Koch, 2003; Zeki, 2003). In fact, it is only a part of the content of consciousness. Similarly, the arousal, the intention and the emotion of consciousness all include their own types and features. For example, the arousal of consciousness includes the features of the degree of arousal, the intention includes the features of different directions, and the emotion includes the features of different feelings.

Because consciousness is subjective, it is always personal. The conscious activities that connect with individual self are called self-consciousness, including all parts of the four elements of consciousness related to individual self. The self-consciousness not only belongs to individuals, but also involves individual conscious activities, such as the activities of the four elements of consciousness related to individual self.

In self-consciousness, the self includes the individual current existence and activities, the individual experience and plans for the future, the relationships between individuals and their environment, and the relationships between one individual and another. The individual confirms themselves through these experience and activities. The individual has self-consciousness, and thus can know himself and his position in the environment (Tang, 2004).

The integrative view about conscious experience discusses not only the elements of consciousness and their integration, but also the brain basis of conscious elements and the brain mechanism integrated by conscious elements at the system level. The four elements which integrate conscious experiences are all indispensable. That is to say, we cannot experience consciousness without arousal of consciousness, and the experience always has content, intention and emotion.

The overall consciousness is unified. The four elements are not paralleling mechanically, but integrated by their interactions. There are close relations between these four elements. The most basic one is the arousal of consciousness, which supports the other three elements, and in turn, the other three elements influence the arousal of consciousness.

The close relationship between the content of consciousness and the emotion of consciousness is described as follows: the content of consciousness is analyzed and evaluated by the brain; the result of the analysis and evaluation controls the emotions; and the emotions regulate the information processing inside the brain, thus impacting the content of consciousness. The close relationship between the content of consciousness and the intention of consciousness is described as follows: the content of consciousness is analyzed and evaluated by the brain to form intentions, which guide the information acquisition, choice and processing, and thus impacting the content of consciousness. The close relationship between the emotion of consciousness and the intention of consciousness is described as follows: the emotions influence the formation and development of intentions while the intentions make the emotions transfer and change (Tang, 2004).

8.3 Integration for Cognition

Cognition is an important part of mind and an advanced function of human brain. It includes activities of many levels and forms, such as sensation, perception, learning, memory, attention, thinking and language, etc. From the perspective of general integration theory, cognition is the integration of activities with these different forms. In the cognitive process, these activities rely on each other and influence each other. Their interactions are usually not only the interactions of certain two kinds of activities, but also the cross interactions of many activities with different forms, which form the complex network of cognitive activities.

Cognitive neuroscience regards cognition as information processing inside the brain. In information processing, there is integration of all kinds of information. This section investigates cognitive process from the perspective of integration of information processing and conscious activities, namely, we may look at cognition as a course of information processing or as conscious activities in the cognitive process. These two aspects are coupled and integrated.

The cognitive process usually begins with the information acquisition from the external environment. The information is processed, subjectively experienced, understood, evaluated by the brain, and then the brain makes decisions on how to regulate and control the behaviors and react to the environment.

The individuals are stimulated by objective things and produce subjective feelings. These feelings are the subjective experience about the content and features of physical stimulus. Individuals have the subjective feelings about the color red of red objects, auditory feelings about sounds, and painful feelings about the hurt in the body. These subjective feelings are the basic features of conscious activities. The subjective feelings of physical stimulation are the basis of cognition, which are indispensable.

The individual cognition has not only subjective feelings about the physical stimulation from objects, but also the understanding about the meaning of the stimulus. The individual accumulates experiences, explains the subjective feelings and organizes all kinds of relevant information. When there is subjective feeling about the color red, the individual makes his explanations about the meaning of the color; when there is subjective feeling of sounds, the individual makes his explanations about the meaning of the sounds; and when there is subjective feeling of pains, the individual makes his own explanations about the meaning of hurt. This understanding of meaning is an important part of conscious activities. The understanding of the meaning of physical stimulus is part of cognitive content, which is indispensable.

There are evaluations and decisions in the cognitive process. The information processing inside the brain includes the evaluation of information and the decision based on the evaluation. Evaluation and decision making belong to conscious activities. In the cognitive process, individuals generate subjective intentions by understanding the meaning of information and evaluating the related events. These subjective intentions make individuals gain new information and influence the development of the cognitive process.

The decisions made by individuals through evaluation and decision making regulate the organs through the functional system of regulation and control inside the brain. They also react appropriately towards the external environment, generating behaviors and acting on the external objective things. The active regulation on cognitive process by individuals is another important part of the conscious activities, which is also necessary in the cognitive process.

All in all, in a cognitive process, there is not only information processing, but also sensation, understanding, evaluation, decision making and regulation, the integration of information processing, and conscious activities inside the brain. Besides, information exchange, subjective feelings, meaning understanding, decision making, active regulation and output action are all necessary. A simple model of cognitive information processing only pays attention to the information processing inside the brain but ignores the subjective feelings, meaning understanding, event evaluation, decision making and active regulation, so it cannot fully understand cognition.

The book *Unified Framework of Psychology and Cognitive Science* (Tang, 2007) proposed two examples of simple cognitive events to illustrate the integration of information processing and conscious activities: One is listening to a speech and the other is recognizing a pattern. When people listen, the brain conducts complex processing of the discourse information. The listener also has subjective feelings about the sound, image and environment of the speaker, retrieving the semantic knowledge stored in his brain to understand the meaning of the heard speech, and then making decisions and reacting on the speech. If people only discuss the information processing of the cognitive process without discussing the conscious activities in the process, they cannot describe the simple cognitive process of listening to a speech.

When people recognize patterns, there is a processing of visual information and lots of conscious activities inside the brain. Simply retrieving the information of figures cannot deeply understand the content of the figures. Understanding the meaning of information, analyzing the knowledge stored in one's memory and speculating are also needed. These top-down processes are all conscious activities. Using a simple model of cognitive information processing to explain the visual pattern recognition is useful but limited. If we only discuss the information processing in the cognitive process without discussing the conscious activities, we cannot completely describe this simple cognitive process.

To sum up, information processing in the brain is an aspect of cognition and the conscious activity is also the important part of cognition. In cognitive process, information processing and conscious activities are closely connected. They are integrated instead of being independent. We have expanded the traditional model of cognitive information processing and proposed the coupling model of cognitive information processing and conscious activities. Cognition is the result of integrating information processing and conscious activities. This model emphasizes the function of conscious activities in the cognitive process on the basis of traditional model of information processing, discussing both information processing and conscious activities, and their coupling and interaction.

The two models are different in that the traditional model does not discuss the subjective conscious activities. In the traditional model, there are only concepts of information and information processing without consciousness and conscious activities. The latter model emphasizes the importance of consciousness and the interaction between conscious activities and information processing. In this model, there are concepts of information, information processing, consciousness, conscious activities, and the interaction and coupling between information processing and conscious activities (Tang, 2007).

8.4 Mental Integration

The last two sections have discussed the integration of consciousness and cognition, both of which belong to mental phenomena. Psychological activities include mind and behaviors. This section will discuss mental integration, and the next section will discuss the integration of mind and behavior.

Mind is the function of the brain with complex structures, including arousal, cognition, emotion, volition and so on. The arousal component is closely related to consciousness. Cognition is an important component of mentality, including many different processes. Mentality is the result of integration of all kinds of mental components.

The mental integration involves very extensive content. The integration of cognition and the integration of consciousness are both a part of mental integration. So is the integration of mental ability. The mental ability is an important issue in psychology. This section will mainly introduce two kinds of intelligence models: the PASS (short for planning, attention, simultaneous and successive) model and the AMPLE (short for attention, manipulation, planning, learning and evaluation) model, and illustrates the mental integration from the perspective of integration of mental ability.

Naglieri and Das (1990), Das, Naglieri and Kirby (1994) explained the features of intelligence from three different levels of cognition, putting forward an intelligence model composed of four processes: planning, attention, simultaneous and successive processing, in short the PASS intelligence model. This model is the cognitive model of intelligence, and discussed in the book *Assessment of Cognitive Processes—The PASS Theory of Intelligence* (Das, Naglieri, & Kirby, 1994) in detail.

The brain science has a profound influence on the intelligence research. The PASS model is a kind of intelligence model based on the achievement of brain science. The basis of this model is Luria's (1966; 1973) theory at three functional systems in the brain, which has been introduced in Chapter 1.

In the PASS model, the four processes of intelligence are based on the cognitive system at three levels, namely the attention-arousal system, the information processing system and the planning system. The attention-arousal system is the basis of the whole cognitive system in this model, the information processing system is in the middle, the simultaneous processing and successive processing are the functions of information processing system, and the planning system is on the highest level in the whole cognitive system. The three systems closely connect with one another, and coordinate to ensure the intelligent behaviors.

Das has designed intelligence scale corresponding with the PASS model, including the four sub-scales which respectively measure planning, attention, simultaneous processing and successive processing. This scale is called the Das-Naglieri cognition-evaluation system (Das, Naglieri, & Kirby, 1994).

We believe that the mental ability is the results of the integration of arousal, cognition, emotion, volition and so on. The ability discussed by the PASS model is only one part of mental ability. Thus according to new experimental facts and the knowledge about brain functional systems, the PASS model can be expanded to the AMPLE model which includes descriptions of arousal, cognition, emotion and volition. The term AMPLE is short for attention, manipulation, planning, learning and evaluation (Tang, 2008b).

The theory of four functional systems in the brain (Tang & Huang, 2003) is the expansion of the theory of three functional systems in the brain (Luria, 1973). According to this theory, human behaviors and psychological activities are realized by the interaction and coordination of four functional systems in the brain. In the intelligence research, to use the idea of four functional systems in the brain to study intelligence will naturally result in the expansion of the PASS model which is based on the theory of three functional systems in the brain. We expand the PASS model on the following aspects (Tang, 2008b):

First of all, the evaluation-emotion process is regarded as one basic process of intelligence. In the PASS model, the concept of evaluation is mentioned, but the evaluation-emotion function is not discussed as an important function of the brain, and the evaluation-emotion process is also not regarded as one basic process of intelligence. The theory of four functional systems includes the fourth system, namely the evaluation-emotion system. The evaluation-emotion process is emphasized by expanding the PASS model. In an individual brain, there is an innate evaluation-emotion structure. The result of evaluation of information will cause the experience of emotion. The process of evaluation and decision making is an important part of intelligence. Emotion is also vital to intelligence.

Secondly, learning and memory are also listed as important processes of intelligence. The PASS model mentions short-term and long-term memory in discussing information encoding, but does not treat them as the basic processes of intelligence. Because learning and memory are important for cognition, the process of learning and memory in the cognitive processing activities should be emphasized in the expanded model.

Thirdly, the content of information encoding and processing in the PASS model is modified. We take the process of manipulating representations as the main process instead of the simultaneous and successive processing in the PASS model. In fact, the process of manipulating representations includes the functions of simultaneous and successive processing.

Fourthly, it is emphasized that mind is integrated by arousal, cognition, emotion and volition. The mental ability is the integration of arousal-attention ability, cognition ability, emotion ability and volition ability. In the expanded model, attention, manipulation, planning, learning and evaluation are integrated as a whole content of intelligence.

To expand the PASS model does not discard the former model but increases new contents while keeping the characteristic contents of the former model, namely keeping the processes of planning and attention while emphasizing the process of evaluation-emotion and learning-memory. The PASS model which belongs to the cognitive model of intelligence mainly discusses the cognitive process in mental activities. The expanded model includes not only cognition, but also emotion and volition, being regarded as an integrated model of arousal, cognition, emotion and volition. The AMPLE model is based on the theory of four functional systems of the brain, believing that the intelligence activities include the following five processes, namely, attention, manipulation, planning, learning-memory and evaluation-emotion.

The first process is attention. The PASS model has already pointed out that the attention arousal is an important process of intelligence. The intelligence activities need the arousal state of individuals to control the effective work of the brain. The attention arousal is the function of the first functional system in the brain, and the related brain regions are brainstem reticular formation and the limbic system. The attention-arousal system is the basis of intelligence activities. The interactions between the attention-arousal process and the other processes are realized by the interactions between the first functional system and other functional systems.

The second process is manipulation. In the brain, the psychological representation of information and the psychological manipulation of representation exist side by side. Information processing inside the brain includes not only the encoding of information, but also the processing of information and the understanding of the meaning of information. These processes are all important contents of intelligence activities. The manipulation of psychological representations includes simultaneous and successive information processing. The process of manipulation is the function of the second functional system in the brain. The related brain regions are occipital lobe, temporal lobe, parietal lobe, and so on.

The third process is planning. The former PASS model has already stressed that planning is an important intelligence process. In the intelligence activities, individuals have to constantly predict and plan. Planning is the function of the third functional system in the brain. Its related brain region is the frontal lobe. The planning system integrates with other systems to coordinate the intelligent activities: planning system promotes or restrains the attention system, monitors and regulates the system which manipulates the psychological representations, plans and adjusts the behaviors.

The fourth process is the learning-memory process. Learning and memorizing are two important intelligence processes. Learning is the constant construction of cognitive structure during the interaction with environment. Intelligence includes the individual's learning ability in the environment. The process of memorizing is the process of encoding, transferring, storing and retrieving the information. There are long-term memory and short-term memory. The process of learning and memory is the joint function of the second and third functional system in the brain. Learning and memorizing integrate with other processes to form the integral intelligence. Without learning and memory, there is no intelligence.

The fifth process is evaluation and emotion. Evaluation and

emotion are also important processes of intelligence. The intelligence activities need constant evaluation and choice making about all kinds of information. The result of evaluation leads to the experience of emotions. This process is the function of the fourth functional system in the brain. Its related brain regions are amygdala, limbic system and part of the prefrontal lobe. The evaluation-emotion process not only influences the planning system and the learning-memory system, but also influences the attention-arousal system and the manipulation of psychological representations.

The aforementioned five processes compose the AMPLE intelligence model. This model suggests that human intelligence is multivariate, and the five processes and their interactions integrate into the whole intelligence. These processes are based on the four functional systems in the brain and coordinate through the interactions between them. This model is based on the four functional systems, with basis of neuro-anatomy and neuro-physiology, thus it is empirical instead of speculative.

The PASS model has inspired many intelligence tests. The AMPLE model is the expansion of the PASS model, so it keeps the former tests of the PASS model and increases a series of new content, including the test of learning, memory, evaluation and emotion. These tests combine with the cognition-evaluation system to construct a more comprehensive intelligence test.

In the field of intelligence, scholars have proposed many kinds of theories. Since the middle of the 1980s, apart from the PASS intelligence model, the representative theories are Gardner's multiple intelligence theory, Sternberg's triarchic theory of intelligence, Salovey, Mayer and Goleman's emotional intelligence theory, and Hawkins and Blakeslee's theories on intelligence. The AMPLE model has something in common with but many differences from these theories on intelligence.

Gardner's (1993) multiple intelligence theory suggests that intelligence is multiple, such as language, logic-math, space, music, body movement, sociality and self-knowledge intelligence. This theory does not discuss the relationship between these intelligences. The AMPLE model discusses different contents and also stresses the integration of multiple intelligence processes and their brain mechanisms.

Sternberg's (1985) triarchic theory of intelligence includes components of intelligence, intelligence situations and intelligence experience. The components of intelligence sub-theory believe that intelligence has three components, namely meta-component, manipulative component and knowledge gaining component. It is the cognitive model of intelligence without discussing emotions and the brain mechanisms of intelligence. The AMPLE model is based on the four functional system theory of the brain. It discusses the cognitive function of the second functional system in the brain, the arousal function of the first functional system in the brain, the intention function of the third functional system in the brain and the evaluation-emotion function of the fourth functional system. Thus the AMPLE model is an integrated model with the components of arousal, cognition, emotion and volition.

Salovey, Mayer (1990) and Goleman's (1995) emotional intelligence theory discusses emotional intelligence specifically, namely the intelligence related with understanding, control and using emotions. However, it does not consider the cognitive process, and volitional process, thus it is not comprehensive. The AMPLE model stresses attention, manipulation, planning, learning, memory, evaluation and emotion.

Hawkins and Blakeslee's (2004) intelligence model emphasizes that the element of intelligence is memory and prediction. The AMPLE model includes the memorizing process and prediction-planning processes. It also points out that there are other intelligence processes apart from memory and prediction. These processes are integrated to form the integral intelligence.

8.5 Integration of Mind and Behavior

Psychological activities include mind and behaviors. Mind is the function of the brain, which is subjective. Behaviors are humans' actions and reactions, which are external performance. Mind and behaviors are closely connected. A good example in point is the relationship between decision making and actions. Decision making is an internal mental activity while behaviors are the explicit activities controlled by the decision. Right decisions cause right action and the results of the action have an impact on the decision-making process.

From the view of general integration theory, human's psychological activities are integrated by mind and behaviors. The unification of mind and behaviors is not the simple sum of the two. They interact with each other and are mingled with each other. The book *On Intelligence—Integration of Mental and Behavioral Abilities* (Tang, 2010) discusses the nature of intelligence and points out that intelligence includes internal mental ability and external behavioral ability, which are different but connected. The mental ability is the feature of mental activities, which defines what can be done by the mental activities. The behavioral ability is the feature of behavioral activities, which defines what can be done by the behavioral activities.

In the book *Human Intelligence* (Pan et al., 1985), Pan pointed out that intelligence includes wit and ability. Wit is the cognitive ability of humans towards objects while ability is the action capability, including skills and habits. Wit and ability cannot be separated. The cognitive ability can be inside the brain with the subjective form, which is wit, but when realized in human's actions, it is ability.

We proposed a general definition of intelligence, regarding intelligence as the integration of mental abilities and behavioral abilities (Tang, 2010). Mental ability is stored inside the brain with subjective form, including attention-arousal ability, cognition, emotion and volition abilities, but it is not confined to the cognitive ability. Behavioral ability is equivalent to the ability realized in human's actions, including the abilities of manipulation, expression, management, sociality and so on. The mental abilities and behavioral abilities are integrated as the whole intelligence.

On the aspect of the intelligence model, the previous section introduces the PASS model and the AMPLE model. The AMPLE model puts forward that the intelligence activities are integrated mental activities, and thus it is a model including arousal, cognition, emotion and volition. Although the AMPLE model has already expanded the cognitive model to become psycho-cognitive model of intelligence, it only discusses the mental abilities while not including behavioral abilities, so it is not complete. In fact, in the intelligence process, the mental and behavioral processes are closely connected and integrated. Chapter 11 will discuss the intelligence integration, emphasizing that intelligence is integrated by mental abilities and behavioral abilities, thus expand the model to one with both mental and behavioral activities, which should be complete and comprehensive.

This chapter has discussed many aspects of integration phenomena in mental activities, such as psychological interactions, consciousness, cognition, mentality, mentality, and behavior. From these discussions, we can find that the integration phenomena in the psychological world have many features that do not exist in the physical world or the biological world.

Chapter 9　Knowledge Integratics

In the long-term practice and studies, human beings constantly gain knowledge of nature, engineering techniques and human society. Through production practice, engineering practice and social practice, especially experimental and theoretical research on natural phenomena by contemporary science, a large amount of new knowledge has been gained, which makes the sum of human knowledge grow by leaps and bounds in unprecedented breadth and depth (Wilson, 1998).

The treasury of human knowledge is integrated by all the knowledge in human history. In thousands of years, the intelligent and hardworking Chinese people have contributed a lot to human civilization. Chinese culture is one part of the treasury of human knowledge.

During the development of knowledge, there exist phenomena of knowledge integration, whose features need special study and whose experience and principles need summarizing with the view of general integration theory. Therefore, we propose building a discipline termed as knowledge integratics, which studies all kinds of integration phenomena of knowledge integration, their principles and applications.

This chapter will first discuss knowledge integration and cross-disciplinary study, and then use a few concrete examples to exemplify several knowledge integration phenomena with different forms, including the integrative model of selective attention, the integration of system of psychological disciplines and the integration of theories of cognitive science.

9.1 Knowledge Integration and Cross-disciplinary Studies

The structure of knowledge is complex and hierarchical. At each level, there are various types of knowledge integration phenomena. The following part will illustrate several aspects of knowledge integration, such as the integration of knowledge resources, theoretical models, disciplinary systems, research approaches and different disciplines, etc.

The integration of knowledge resources is a kind of knowledge integration. As far as knowledge integration is concerned, people are most familiar with libraries and reference rooms, including digital libraries and electronic reference rooms developed in these years. We do not elaborate on them here, but focus on internet-based science databases and data sharing. In some fields of contemporary science and technology, there are many departments working in the same field. The most common one is diagnosis and treatment of clinical medicine. People accumulate a large number of scientific materials in the long-term practice. If each department works separately, the materials they have will be dispersive and repetitive, but in fact, many of them complement each other. Thus we need to collect these materials, categorize and save them.

Currently, many fields of science and technology are engaged in the integration of knowledge resources, collecting science materials related with their own field, and building scientific database for researchers from different organizations. These resources are shared between researchers. We can obtain data from different departments through scientific database to conduct scientific study and practical applications, which is called the sharing of scientific data. Scientific database includes not only all kinds of detailed scientific data, but also expert knowledge, such as their experiences and achievements, and scientific tools, such as the pragmatic analyzing software and graphs.

The integration of scientific data, research achievements, expert knowledge and scientific tools mentioned here all belong to the integration of knowledge resources. How to integrate useful scientific data, research achievement, expert knowledge and scientific tools, and how to build intellectual scientific database and provide efficient sharing of scientific data for society are the research subjects of knowledge integratics.

The integration of theoretical models is a kind of knowledge integration. Building a model is a way of approximate description of entities. When we study an object, we usually propose some theoretical models to explain its features.

As far as more complex objects are concerned, they have many different aspects. In the process of understanding complex objects, we should observe different aspects of them from different perspectives, proposing different models and explaining different features under different conditions.

Thus, at the first stage of study, there might be many different theoretical models towards relatively complex objects, approximately describing their different aspects. Then, due to the development of understanding, the advantages of each model can be gradually integrated, and a unified theoretical model can be built, which is the integration of theoretical models. For instance, in the study of nucleus, the water drop model and the shell model were proposed before the unified model of nucleus.

In section 9.2, the study on the phenomenon of selective attention and the integration of theoretical models of selective attention will be discussed.

The integration of disciplinary system is a kind of knowledge integration. A big discipline usually includes many small sub-disciplines. At the beginning stage of disciplinary development, there is no unified disciplinary system. These scattered sub-disciplines are independent but actually connected. In the process of development, the relations of these sub-disciplines are gradually known to us. Eventually, a systematic disciplinary system will be built under the unified framework based on these various sub-disciplines. In section 9.3, the structure of psychological disciplines and their integration will be discussed.

Another kind of knowledge integration is the integration of different research approaches in the same field. In the developing process of a research field, people have different viewpoints about research, thus different academic ideas and research trends emerge. Those thoughts that influence the development of the field are summarized as the research approach in this field. In an immature research field, there are usually many different research approaches. As this research field develops, the relevant knowledge accumulates to such a degree that it is abundant enough to integrate various research approaches to form a unified research approach. This unified research approach absorbs the essence of each kind of separate research approach. Section 9.4 will discuss the different research approaches in the field of cognitive science, and the integration of research approaches in cognitive science.

Next, cross-disciplinary study, kind of knowledge integration, will be discussed. Chapter 4 has already mentioned the integration phenomena of cross-disciplines. In contemporary natural science and engineering technology, many scientific and engineering problems involve various disciplines, especially many emerging disciplines, such as nano science and technology, biological science and technology, information science and technology, and cognitive science and technology, etc. Under many circumstances, only relying on one kind of discipline cannot solve the problem. By integrating the knowledge and technology of different disciplines can the problem be solved. This calls on the issue of cross-disciplinary study, which is the integration of different disciplines.

Take astronomy as an example, which is developed by cross-disciplinary studies. Traditional astronomy emphasizes the use of optical methods, such as optical imaging and spectral analysis method to observe celestial bodies and celestial phenomena. Many achievements have been gained in this field, especially in the studies of celestial appearance, spatial distribution of celestial bodies, celestial movement, and celestial interactions at the cosmoscopic level. Modern astronomy measures nucleus and particles in the space besides optical observations. Theoretically, we study not only celestial phenomena at the cosmoscopic level, but also the celestial phenomena, including the processes in the celestial bodies and their evolution at the microscopic level. The nuclear physics, particle physics and traditional astronomy combine with each other to generate nuclear astrophysics, particle astrophysics, etc.

Another example is the cross-disciplinary study between neuroscience and other disciplines. Scientific problems of many disciplines involve the activities of human brain and mind, and the interactions between human beings and their environment, which are closely related to neuroscience. The development of modern neuroscience offers these disciplines many chances to engage in new research. The integration of neuroscience and these disciplines generate large amounts of cross-disciplines.

Among them, people are most familiar with the following cross-disciplines-neurochemistry, neuro-endocrine-immunology, neuropsychology, cognitive neuroscience, etc. Neurochemistry is the cross-discipline of neuroscience and chemistry; neuro-endocrine-immunology is the cross-discipline of neuroscience, endocrine and immunology; neuropsychology is the cross-discipline of neuroscience and psychology; and cognitive neuroscience is the cross-discipline of neuroscience and cognitive science.

In recent years, many new cross-disciplines are emerging, such as neuroinformatics, neurolinguistics, neuroeconomics, neuromanagement, educational neuroscience, etc.

For instance, neuroinformatics is the cross-discipline of neuroscience and informatics. The fruits of neuroscience inspire and promote the development of informatics, while the results of informatics provide new concepts and research tools for the research of neuroscience. Neuroinformatics use the concept of information and information processing to study the information problem of nervous system, such as the carrier of nervous information and its production, encoding, storage, retrieval, transportation and processing. These studies in turn enlighten the machine intelligence. This discipline also exploits modern information tools and technology to build databases of neuroscience at different levels and to provide a platform for data sharing of neuroscience research. Thus, we can rely on it to analyze large amounts of neuroscience data, to build models and to conduct theoretical studies.

Besides these two disciplines in the cross-disciplinary studies of neuroscience and informatics, there are many other disciplines, such as biology, medicine, mathematics, physics, chemistry, computer science and engineering technology science, etc., including lots of knowledge integration and technology integration.

Neurolinguistics is the cross-discipline of neuroscience and linguistics. Neuroscience studies the neural basis of linguistic phenomena to promote the development of linguistics, and the phenomena of linguistics enlighten neuroscience and provide new research themes.

The integration of arts and science is mainly the integration of relevant knowledge of arts and science. The development of modern society has brought up large amounts of complex problems of science and technology, such as environment, energy, climate, population and health, mineral exploitation, safe production, protection of biological diversity, and sustainable development of economy, which are originated in the problems of natural science and engineering technology, and related to various social factors, exerting wide social influence. Thus, they need the cooperative study by natural science, engineering technology, humanities and social science, which leads to the cross-disciplinary studies of relevant fields.

For example, modern economy develops as a cross-discipline. In the past, economics conducted mostly qualitative or half quantitative analysis and studies. With mathematics providing us with more and more research tools, economy combines with mathematics, integrating qualitative and quantitative research to generate the cross-discipline as mathematical economics.

Other examples are neuroeconomics and neuromanagement. Neuroeconomics is the cross-discipline of neuroscience and economy. On one hand, neuroscience studies the neural basis of cognitive process in the economic activities, enlarging the understandings of social cognitive mechanism; on the other hand, the research results promote the development of economy. Neuromanagement is the cross-discipline of neuroscience and management. The studies on management by neuroscience enrich the knowledge of neural science, and provide new scientific basis for management science as well.

The complex problems of science and technology need joint research of different disciplines from different perspectives, with different methods and based on different experience. The knowledge and technology of different disciplines are complementary, which can be integrated through cross-disciplinary study to solve the common problems.

Cross-disciplinary study should be substantial instead of superficially putting several different disciplines together. Integration is needed to solve common scientific problems, concepts and methods. Besides the integration of knowledge and technology, we need to organize the support and resources of different disciplines to cooperate with each other. If we only put several disciplines together, divide a research project into unrelated parts with staffs of each discipline doing their own jobs, and then pool the results, we cannot solve the problem. This is not a real cross-disciplinary study. In order to conduct knowledge integration of several disciplines efficiently, we need to summarize the experiences of cross-disciplinary research.

Cross-disciplinary study is a process of long-term knowledge integration. Knowledge needs constant accumulation, instead of being built in one day. Thus a complex cross-disciplinary scientific problem is usually divided into several stages. Knowledge integration is also a process of exploration. In the long-term practice, we discover knowledge, summarize experience, correct our errors to make new achievements. The process of knowledge integration not only organizes the existing knowledge of several aspects, but also generates new concepts and problems, initiating new hypotheses and plans which are examined by new experiments to improve the conventional understandings. This is called knowledge innovation.

9.2 Integrative Model of Selective Attention

This section introduces the integrative model of selective attention as an example of integration of theoretical models. Selective attention is an important function of the brain. In conscious mental activities, selective attention plays an important part. Various conscious activities in the brain implicate selective attention.

Corbetta et al. (1991) once conducted experiments with the technique of PET brain imaging. In the experiments, the participants were shown visual images and asked to report the shape, color, and change of speed of the image. When the participants passively perceived the stimulus, some specific areas of the brain were activated. But when the participants noticed the stimulus positively, the activation level of the relevant brain areas increased, and was even accompanied by the activation of basal ganglion and brain areas in anterior cingulate gyrus. Hillyard and Picton (1987) once used the technique of event-related potentials (ERP) to conduct an experiment of selective attention, showing that the participants' ERP signals increased when they focused on the stimulus. Duncan, Ward and Shapiro (1994) observed in their experiments and found that within several hundred seconds after the participants were attentive to the given stimulus, they could hardly react to the next item. One feature of selective attention is that only one item can be noticed at a time.

To account for the experimental facts of selective attention, we proposed the integrative model of selective attention, namely the discrimination and coincidence model of selective attention, believing that the mechanism of selective attention is like the electronic circuits with the functions of discrimination and coincidence (Tang & Guo, 2000). The discriminator circuit and the coincidence circuit are the standard circuits in nuclear electronics. They are usually made into modules, which compose the complex electronic system.

Chapter 4 introduces the discriminator circuit, which selectively records the events if the amplitude of input signal exceeds a certain level. This level is called the discrimination level. Chapter 5 introduces the coincidence circuit, which selectively records the simultaneous input signals. The coincidence circuit might be a two-fold coincidence circuit if there are two inputs or a multiple coincidence circuit if there are more than two inputs. One kind of input signals can be the control signal, namely the gating signal.

In our integration model, the signals of different items are the input signals for the circuit, which enter each discriminator circuit and pass them to the coincidence circuit. The attention control signal is the gating signal to the coincidence circuit. The joint circuit composed of discriminator circuit and coincidence circuit selects the following event: the amplitude of the input signal exceeds a certain level and occurs simultaneously with the attention control signal.

According to this model, many input signals are processed in many parallel input pathways of the circuit. There is a discriminator circuit in each pathway. If the amplitude of processed signals exceeds the level of discrimination, that pathway produces an output signal. It is also required that the discrimination level of the parallel input pathways in the circuit can be automatically adjusted to make sure that only in one pathway can the input signal exceed the discrimination level. Thus, this pathway can produce an output signal while other pathways do not.

Then, the output signal is transported to the two-fold coincidence circuit as one input, and the other input is the attention control signal. In the two-fold coincidence circuit, only when the two signals input within the resolution time of the coincidence circuit, there will be coincidence output, but if the two signals input beyond the resolution time, there will be no coincidence output. To summarize, the output signals of discriminator circuits, which are produced by the input simultaneously with the attention control signal within the resolution time of the coincidence circuit, will be recorded in the coincidence circuit (Tang & Guo, 2000).

Scientists have proposed many models of mechanism of selective attention, such as the filter model of selective attention (Haberlandt, 1997), the spotlight model of selective attention (Crick, 1984), and the biased competition model of selective attention (Desimone & Ducan, 1995). According to the filter model, the mechanism of selective attention is like a filter. Based on the spotlight model, the mechanism of selective attention is like a spotlight of mind, intensifying the processing of illuminated items. The biased competition model suggests that the mechanism of selective attention is like a competition among all the processed items, for instance, the entities in our sight which will compete for visual attention to represent, analyze or control. These models can not only describe different aspects of selective attention, but also have their disadvantages.

The discrimination and coincidence model of selective attention is actually the integrative model of selective attention. It integrates every feature of selective attention, and at the same time combines the former models, especially the searchlight model and the biased competition model to form the integrative model of selective attention.

According to the integrative model of selective attention, the inputs of different items have different intensities and different top-down influences. They will compete with each other. One of them will gain advantage over the others for its largest intensity, which results in the highest activation level. The function of attention is to intensify the item being noticed, instead of scanning every item. It intensifies the salient item and in the meantime it restrains the other unobtrusive items.

The difference between the integrative model and the other two models lie in the following aspects. In the spotlight model, the noticed items are passively intensified, while in the integrative model, the input items positively compete with each other to be the one that is most intensified. In the biased competition model, it puts emphasis on the competition between different items without stressing the importance of intensifying function of attention, while in the integrative model, both competition and intensification are emphasized (Tang & Guo, 2000).

9.3 Integration in Psychological Science

Today, psychology has been developed as a very prosperous discipline. In this process, not only the subfields in psychology itself have been deeply studied, but also a lot of cross-disciplines have emerged during the cross-studies of psychology with other disciplines.

Currently, psychology has many branches, such as cognitive psychology, thinking psychology, psycholinguistics, emotional psychology, neuropsychology, physiological psychology, biopsychology, health psychology, medical psychology, clinical psychology, evolutionary psychology, developmental psychology, educational psychology, school psychology, sports psychology, personality psychology, psychophysics, engineering psychology, environmental psychology, social psychology, industrial and organizational psychology, career psychology, management psychology, advertising psychology, judicial psychology, and military psychology, to name a few.

Some scientists have discussed the paradigms of psychology. According to the opinion of Kuhn (1970), the scientific paradigm of a specific discipline is about the content of this discipline, and it is a kind of world view. He held that the scientific paradigm is the conceptual framework of a discipline, combining with the theories and methods and a series of commonly approved research principles. Hilgard (1987) believed that current psychology is still at the stage of pre-paradigm because there is no single advantageous theory and method in psychology.

Faced with the huge number of various branches of psychology, it needs a unified disciplinary system. Staats (1999), Sternberg and Grigorenko (2001), Denmark and Krauss (2005), and Sternberg (2005) expressed their expectations for the unified theory of psychology. A unified disciplinary system will contribute to the further development of psychology.

In the book *Unified Framework of Psychology and Cognitive Science* (Tang, 2007), based on the theory of grand unification of mental interactions, the theoretical framework of grand unification of psychology was built, which was mainly concerned with the integration theory of psychology science. From the view of mental interactions, the current different branches of psychology actually study different mental interactions. For example, biopsychology and psychophysiology both study the interactions between mind and brain, and between mind and body. Biopsychology emphasizes the brain's biological activities and the influence of other biological activities on mental processes, while psychophysiology stresses the influence of mental processes on the biological function of the brain and the other biological functions of the body.

Because different mental interactions can be unified, different psychological research areas can also be unified. The grand unification of psychology faces the whole picture of psychological phenomena, covering every research area in psychology. Based on the grand unification theory of psychological interactions, the grand unification theory of psychology examines the relations of each research field, studies their integrations and develops the cross-studies of each field from the perspective of psychology and its actual applications on various aspects according to the principles of unification of mental interactions.

The main content of the grand unification theory of psychology includes the viewpoint of mental interactions, the unification of these interactions, the belief that psychology should be unified, and the opinion of integration of different branches in psychological science. This theory is called grand unification theory of psychology as it is constructed based on the grand unification theory of mental interactions. Among them, "grand unification" refers to the unifications of various kinds of mental interactions, and to the idea that guided by this theory, each research field and branch can be integrated to form a unified disciplinary system in psychology.

The book *Unified Framework of Psychology and Cognitive Science* (Tang, 2007) discusses the integrations of the branches in contemporary psychology, which integrates the branches into a unified disciplinary system on the aspects of concepts and methods. Among branches of contemporary psychology, some branches emphasize the basic principle of psychological activities, some others develop in the cross studies between psychology and other disciplines concerning the same issues of these disciplines, and still some other branches emphasize the real applications of psychology in relevant fields.

The unification of different mental interactions grants the inner relations among different branches of contemporary psychology with different contents, and thus the unified disciplinary system of psychology can integrate different branches together. In the following, we will discuss and classify each psychological branch from the perspectives of interactions of components in mental activities, mind-brain interaction, mind-body interaction, mind-matter interaction, mind-society interaction, etc. These psychological branches can be categorized into four types according to the mental interactions they are concerned with.

The first type is the psychological branches concerned with basic psychological traits and processes, and the mental interactions involved are the interactions of components of mental activities and mind-brain interaction. These branches include neuropsychology, cognitive psychology, mental psychology, thinking psychology, psycholinguistics, emotion psychology, personality psychology, etc.

The second type is the psychological branches which are related with many aspects of human body. These branches mainly study mind-body interaction, and also involving mind-brain interaction, etc. These branches include biological psychology, physiological psychology, psychophysiology, health psychology, medical psychology, clinical psychology, and rehabilitation psychology, which are relevant with mind-body interaction.

The third type is the psychological branches which are related with environment. These branches mainly study mind-matter interactions, also involving mind-brain interaction and mind-body interaction, etc. These branches include psychophysics, evolutionary psychology, environmental psychology and engineering psychology which are relevant with mind-matter interactions.

The fourth type is the psychological branches which are related to social practice. These branches mainly study mind-society interactions, also involving mind-matter interaction and mind-body interaction, etc. These branches include social psychology and developmental psychology, and also many applied disciplines such as educational psychology, school psychology, industrial and organizational psychology, career psychology, managing psychology, advertisement psychology, judicial psychology, and military psychology, all of which are relevant with mind-society interactions.

Psychology science is composed of these four types of disciplines. In this integrated system, each branch has its own position based on its different mental interactions, meanwhile different branches are related with one another due to the unification of psychological interactions. They are not in chaos or independent of the whole system.

The grand unification theory of psychology, including a psychological-disciplinary system, is based on the long-term achievement of psychology. It integrates these achievements but does not contradict with the contemporary psychology. The grand unification of psychology is not a new discipline, and cannot replace the specific research of each field in psychology. It has proposed a kind of research approach in psychology, namely the unified research approach. The grand unification of psychology does not end the development of current psychology, but opens up the road for further development to promote the prosperity of psychology (Tang, 2007).

The paradigm of psychology is very important for psychological study. The grand unification theory of psychology can be one candidate theory of building the paradigm. The concept of many mental interactions, the concept of unification of mental interactions, and the concept of integration into a unified disciplinary system by different branches in psychology, can provide us with a means of observing mental world and psychology. This unified disciplinary system provides us with a conceptual framework, which can probably unify different theories and methods of contemporary psychology. And it may also act on a kind of theory, which might integrate various disciplines into a unified psychological-disciplinary system.

Because the grand unification theory of psychology studies various kinds of mental interactions and their unification, it can include all theories and methods of mental interaction researches. In this way, Kuhn's (1970) requirement might be met: different theories exist in the same paradigm to solve different problems, while the paradigm provides the conceptual framework for different disciplines.

9.4 Integration of Theories in Cognitive Science

Human cognition is very complex. Cognitive science studies the cognitive phenomena, principles and applications. Cognitive science research is one of the frontiers of contemporary science.

There are many different ideological trends in cognitive studies, each of which becomes a research approach of cognition. Gardner (1985) and Haberlandt (1997) pointed out that there are many research approaches in contemporary cognitive research, such as the research approaches of neurobiology, information processing, embodied cognition, situated cognition, social cognition, evolutionary cognition, developing cognition, and artificial intelligence, etc.

What are the relations between these research approaches? Can they be integrated? This is a valuable topic. Newell (1990) once discussed the unification theory of cognitive science in 1990. He believed that the aim of science is a unification, and that psychology needs a unification theory, as do various cognitive processes. After listing the cognitive processes, such as problem solving, decision making, learning, memorizing, skill, perception, movement, language processing, motivation, emotion, imagination, dream, daydream, etc., he said that there should be a unification theory of cognition which can integrate these processes, and he put forward a SOAR theory as one kind of unification theories. Additionally, some other scholars suggested other theories, for instance, ACT-R theory (Anderson, 1983; Anderson et al., 2004). However, Sternberg (2005) thought that there had not been any satisfying theory yet until the publication of Newell's works due to the complexity of cognition.

The book *Unified Framework of Psychology and Cognitive Science* (Tang, 2007) briefly illustrated different research approaches of cognition in contemporary cognitive science. Let's recall them one by one:

> The research approach of neurobiology exploits neurobiology to study cognitive processes, and discusses the neurobiological basis of cognitive processes. This approach believes that human's cognitive processes are closely related with the neural activities in the brain, and thus it is necessary to know the neural correlates of various cognitive processes. This research approach is concerned with the particular brain region involved in the particular cognitive process, and the neural activities in these processes (Kosslyn & Koenig, 1995; Gazzaniga, 2000).
>
> The research approach of information processing exploits the information processing in the brain to study cognitive processes. This approach believes that human's cognitive process is the process

of encoding, storing, retrieving and manipulating information from external environment. This research approach is concerned with the ways and mechanisms of information processing in the cognition process (Newell & Simon, 1972; Haberlandt, 1997).

The research approach of embodied cognition probes into the mind-body relation to study cognitive processes. This approach holds that cognition is rooted in human body, and manifested in the body. It also believes that the cognitive process involves and relies on human body, and thus cognition is embodied. This approach stresses that the human body affects cognitive processes, and the issue it concerns is the relationship between cognition and body, and the effect of physical factors on cognitive processes (Varela, Thompson, & Rosch, 1991).

The research approach of situated cognition applies the view that cognition is correlated with the situation to study cognitive processes. Situated cognition means that cognitive process happens in real environment and relies on real situation. It emphasizes that a cognition process should be rooted in situations, and thus the cognitive process is situational. This approach is concerned with the relationship between cognitive processes and real situations (Gibson, 1979; Brooks, 1991).

The research approach of social cognition studies cognitive processes with the view of social environment. Social psychology believes that a human is a social being, and social environment is closely related with the human's cognitive processes. Therefore, the interactions between social environment and individual cognitive process, and the influence of society and culture on individual cognitive process should be studied. This research approach is concerned about the effect of family, group and society on cognitive processes and the cognitive processes in local social situation, and so on (Cacioppo et al., 2002).

The research approach of evolutionary cognition studies cognitive processes with the view of biological evolution. This approach believes that the biological elements, like genetic factors are very important for individual cognitive processes, and it stresses the study of relations between biological evolution and cognitive processes, and between genetic factors and individual cognitive processes.

The research approach of developing cognition studies cognitive processes with the view of individual development. It believes that cognitive ability changes throughout one's life. It stresses that we should study the developing process of individual cognition, including children's cognitive development and the relations between childhood cognition and adult cognition.

The research approach of artificial intelligence studies cognitive processes with the view of comparisons between human brain, computer, and artificial intelligence. It emphasizes the brain-like machine and machinery cognition, etc.

Thus, there are many different research approaches in contemporary cognition research. When he discussed the basic problems of psychology, Glassman (2000) said that one of the fundamental issues of psychology is how to deal with the situation of coexistence of different research approaches. How to deal with the situation of coexistence of different research approaches in cognitive science is also one of the fundamental problems in cognitive science. Confronted with the situation of varying opinions in cognitive science, these different research approaches should be integrated to build a unified theory of cognitive studies. Namely, a unified theory of cognitive science is needed to integrate these different research approaches with different concepts and methods.

Different research approaches in contemporary cognitive science have different emphasis that all have some experimental evidences, research viewpoints and also their shortcomings. From the perspective of various mental interactions involved in cognitive process, they emphasize different mental interactions. Some research approaches are deficient because they only study one kind or some kinds of mental interactions, or they only focus on some aspects of a certain mental interaction without covering all the mental interactions in cognitive processes. However, the concepts of each research approach do not contradict each other, but actually, they are complementary. Each research approach has its own positive and beneficial viewpoints, and each of them is appropriate for discussion about particular mental interactions. Thus we need to exploit these positive and beneficial points and integrate them in the unified research approach. Since each research approach can complement and merge with others, the unified research approach should integrate their positive and beneficial viewpoints.

Neurobiology cares about the neurophysiological basis of the cognition process. The neurobiological process is the physical basis of cognitive activities. To understand the nature of cognition, studies of brain functional activities and physiological activities related with cognitive activities are very important. From the aspect of mental interactions, this research approach emphasizes the mind-brain and mind-body interactions in cognitive process from the perspective of neurobiology. Though many works of this research approach are very beneficial to understanding these interactions in the cognitive process, their shortcomings lie in the lack of discussion about the interactions among all the components of psychological activities during the cognitive processes and the mind-matter and mind-society interactions.

The research approach of information processing examines the information processing in the brain in the cognitive process. During the cognitive activities, information processing and conscious activities are coupled with one another. To exploit the information processing and conscious activities to study the cognitive process is very beneficial for understanding the relations of components in psychological activities and the mind-brain relations. In addition, to study the influence of physical factors on inner psychological activities and brain mechanism is very interesting. From the perspective of psychological activities, this approach prioritizes the relations of the components in psychological activities, mind-brain and mind-matter interactions in the cognitive process. Its feature is involving the inner information processing of many kinds of mental interactions, but it rarely discusses the mind-body interaction in the cognitive process, nor discusses the effect of the physiological processes on cognitive activities.

The research approach of embodied cognition investigates the body in the cognitive process, and the research approach of situated cognition examines the environment of cognitive process. Both research approaches are important for studying cognition. Embodied cognition stresses the relationship between cognition and body. From the perspective of mental interactions, this research approach emphasizes the mind-body interaction. Situated cognition stresses the relationship between cognition and situation. From the perspective of mental interactions, this research approach emphasizes the mind-environment interaction. The following are their shortcomings: they rarely discuss the interactions of various components of psychological activities in the cognitive process, nor do they discuss the mind-brain interaction.

Social psychology cares about the relationship between social environment and individual mentality. Social environment and culture have a big influence on the cognitive process, which is very important for understanding the relationship between cognition and society. From the perspective of mental interactions, this approach mainly involves the mind-society interaction in the cognitive process, and also discusses the influence of the social environment on the interactions of each component of psychological activities and mind-body interaction. Its disadvantage is lack of examination of mind-body interaction and mind-matter interaction in cognitive process.

Other research approaches in cognitive science, like evolutionary psychology, developing psychology and artificial intelligence, also have their own characteristics. For example, evolutionary psychology cares about the effect of biological evolution on psychology and behaviors. From the perspective of mental interactions, this kind of research approach involves various kinds of mental interactions in cognitive activities. To use this approach to study cognitive activities, it is useful to understand the evolving process on various mental interactions in cognitive activities. Another example is developing psychology, whose approach focuses on long-term human life, especially the mentality and behaviors of children and adolescents. From the perspective of mental interactions, this kind of research approach involves many kinds of mental interactions. To use this approach to study cognitive activities is beneficial for understanding the development of each kind of mental interaction.

The integration theory of cognitive science exploits all kinds of mental interactions and the unification to investigate various research approaches in contemporary cognitive science, and integrates the useful viewpoints of these research approaches. This kind of integrating approach is called the unifying approach in cognitive science. Next, we will discuss the concerns, main points, views and methods of this research approach.

The unified research approach cares about all the mental interactions in the cognitive science and their unification. This kind of research approach requires comprehensive investigation and examination of all mental interactions and it holds that different mental interactions have the same base and thus we can conduct unified research on all mental interactions. This kind of approach not only analyzes the features of each kind of mental interaction in the cognitive process, but also investigates the dynamic process of mental interactions.

Although each research approach in contemporary cognitive science emphasizes different kinds of mental interactions of cognitive processes, we prefer the unified research orientation because it can include all kinds of research approaches of cognitive science and integrate all positive and beneficial contents of these approaches and discuss all the mental interactions and their relations and unification. To conclude, the unified research approach is a preferable one because it discusses not only some mental interactions of the cognitive process or some aspects of one kind of mental interaction, but all aspects of mental interactions.

The unified research approach analyzes various research approaches of cognitive science in detail, studies their features, retrieves their useful contents and methods, absorbs the essence, and then integrates them under the unified theoretical framework. Thus its research field includes all the fields of cognitive science, more widely than other research approaches.

The unified research approach believes that contemporary cognitive research approaches all discuss important aspects of cognitive activities. A cognitive process has the basis of neurobiology with the feature of information processing and conscious activities. The cognitive process is embodied, situational and social, and cognition is evolutionary and developmental. The cognitive process involves all kinds of mental interactions with close relations. In cognitive process, there are integration processes of many mental interactions.

Chapter 10　Engineering Integratics

In the field of engineering and technology, there are various integration phenomena. To apply the view of general integration theory into this field and to study the features of integration phenomena in this field are helpful for solving the complex problems in the field of engineering and technology.

In the application of general integration theory into this field, we propose a new discipline, engineering integratics, which studies the integration phenomena, principles and their applications in engineering. We also propose a new discipline, technology integratics, which studies the integration phenomena, principles and their applications in technology.

The research content of engineering and technological integratics is profound. Chapter 4 has mentioned the integration phenomena in the technological field. This chapter discusses several examples of engineering and technological integration, which are the integration of "big science" plan, giant experimental devices, and medical imaging technology.

10.1 Integration in "Big Science"

A large-scale scientific program emerges during the development of contemporary science and technology. These programs have complex scientific problems to solve, which need long-term plan, a large amount of money, organization of many scientific departments and research staff. Compared with a small-scale scientific program undertaken by a team in the laboratory, they are big, thus called "big science" plan or "big science" program. The "big science" plan is a kind of research method of contemporary science and technology, which bears many features different from small-scale programs.

The USA has conducted several "big science" projects, such as the Manhattan Project, Project Apollo and Human Genome Project, all of which are examples of "big science" plans. The Manhattan Project was the construction of the atomic bomb in the 1940s; Project Apollo was the project of landing American astronauts on the moon in the 1960s and 1970s; the Human Genome Project aimed to identify and figure out the sequences of the chemical bases that make up human DNA in the 1990s. Currently, many countries worldwide are conducting the plan of International Thermonuclear Experimental Reactor, which is also an example of "big science" plans. This plan is a project of studying reactor plant of thermonuclear fusion. China has successfully conducted several "big science" projects, and will implement several new ones.

Lambright (2009) discussed the implementation of a "big science" project in the Monograph, *The Key of "Big Science"—Managing and Coordinating*, which introduced several cases of "big science" plan in the USA. In this book, the development and management of the Human Genome Project were illustrated in detail and some funny stories during the implementation of the project were mentioned.

This book introduces the managing work of "big science" projects from the aspects of aim, organization, support, competition, and leadership, pointing out that management and coordination are the keys to the implementation of a "big science" plan. It also discusses some other large scale research and development plans, including nanotechnology and international space station.

From the perspective of general integration theory, many integration phenomena can be found during the planning and implementation of "big science" plans. These big science projects solve the key problems of science and technology, and they also belong to engineering projects. The integration in the "big science" projects is an important topic in engineering integratics.

From the preparation of a "big science" plan to the scheme, proposal, implementation till the accomplishment of the plans, there are many forms of integration processes at different stages. As Lambright (2002) mentioned, management and coordination are important content of integration phenomena. The organization and the management of a "big science" project are all integration process, such as the integration of all kinds of scientific thoughts and research program, integration of multiple subjects and technologies, integration of different scientific departments and teams, and so on. A large amount of coordinating work should be done during the integration processes to guarantee that the "big science" project is accomplished efficiently and effectively.

In fact, integration processes not only exist in "big science" projects, but also in all kinds of scientific research. In former cases, integration processes are larger scale and more of an obvious manifestation and play a more important role. In science and research integration, these features and principles of integration processes should be studied, and the implementation of scientific plans needs to be guided by these features and principles.

10.2 Integration of Large Scale Experimental Facilities

The integrations in the design, construction and operation of large scale experimental facilities are examples of engineering and technology integrations. Contemporary scientific experiments not only conduct various small scale experiments in many laboratories, but also build some big scientific experiment centers, equipped with many large scale experimental facilities, which can be simultaneously used by many departments to conduct multiple scientific experiments.

Large scale experimental facilities belong to complex engineering infrastructure, including many engineering systems, which are integrated into a large scale experimental facility; each engineering system is further composed of many experimental devices, which are integrated into an experimental system with complete functions.

For example, the large-scale facilities in the experimental center of synchrotron radiation light source include the devices for generating synchrotron radiation light and many experimental stations. The devices which generate synchrotron radiation light include electron accelerators and other devices for stably making high-energy electron beam work. Chapter 5 mentioned synchrotron radiation. High-energy electron beam travels around the ring-shaped devices, generating radiation light from the pipes along the tangent lines of the circles, then the light is sent to many experimental stations. Every experimental station is equipped with various targets and specific detection equipment, such as electron spectrometer for measuring the properties of the produced electrons generated by the hit of synchrotron radiation on the targets.

Large scale experimental facilities of high-energy physics experimental center have all kinds of systems of accelerators and detectors. Among them, the system of accelerators speeds up the electric particles, providing the particle beams for high-energy physical experiment; the detection system has many specialized kinds of equipment to measure the properties of high-energy reaction products.

To illustrate the complex and integrative nature of this kind of large scale experimental facility, large electron-positron collider (LEP) in the European Organization for Nuclear Research (CERN) is an example (Tang, 1985; 1993). Positron and electron collide in the device, generating a total energy of 200 billion electron volts. The collider in this energy range can produce many intermediate boson particles Z^0 and $W^{\pm}$, thus LEP is called the factory of intermediate bosons.

The main body of this collider was built in the boundary of France and Switzerland. The main ring and several experimental halls of collision regions were underground, 50 to 170 meters deep underneath. The perimeter of the main ring was 27,000 meters. The collider mainly includes the devices of accelerating electrons, the devices of generating and accelerating positrons, and the collision areas, etc. The positrons and electrons move in opposite directions, accomplishing the collision in the four collision areas of the main ring. This collider began construction in 1983, and was completed and put into use in 1989.

At four collision areas, four large scale detectors were installed, including OPAL detector, ALEPH detector, L3 detector and DELPHI detector, measuring the products of high-energy positron-electron collisions. Every detector system is a huge experimental device. For example, L3 detector is a large scale detector with a solid angle of

4π and large-volume magnetic field. The feature of this detector is to precisely measure photons, electrons and muons. The total weight of the detector is 8000 tons. In the volume of $12\times12\times14$ m^3 space, the ordinary magnet produces a uniform magnetic field with the intensity of 0.5 Tesla (Adeva et al., 1990).

Looking outside from the collision area, L3 detector is composed of the following layers. (1) A vertex chamber is around the collision area, called time expansion chamber, outside of which there are four layers of proportional chambers measuring the horizontal coordinates of particles. (2) An electromagnetic calorimeter is composed of bismuth germanate crystals outside of the vertex chamber. The collision area was surrounded by a large number of strip-shape bismuth germinates. The calorimeter can measure photons and electrons of more than 2 billion electron volts with the energy resolution higher than 1% according to the combined information given by the vertex chamber and electromagnetic calorimeter, photons and electrons. (3) A hadron calorimeter is laid further outside. There are 144 modules in the barrels of hadron calorimeters circling the beam line, and also end-cap part of hadron calorimeter covering the two ends. They are composed of uranium plates and proportional chambers, which can measure the energy and position of incident hadrons. (4) A large scale muon drift chamber can measure the muons with momentum resolution of 2%, and additionally, the beam luminosity monitors are put 2.75 meters away on both sides of the collision point.

Many interesting physical results have been achieved by L3 detector, such as the measurements of the properties of Z^0 and $W^{\pm}$ particles, the determination of neutrino species, the measurement of parameters of electroweak interactions, the characteristic of decay of Z^0 into b quark, the test of quantum chromodynamics, the determination of strong coupling constant, the test of quantum electrodynamics and the measurement of lepton properties, etc. (Tang, 1993).

The integration of large scale experimental facilities is an important topic of engineering integratics. The aim of building these facilities is to conduct scientific experiments with these facilities. The efficacy of the facilities is represented by advanced design, good operation, stable work, effective use of machine time, and large numbers of high quality experimental achievements. Each system and device in the facilities should be integrated correctly and efficiently. Their coordinating and stable work is very important for raising the utilization of devices and production rate of experiments.

The optimization of integration from planning, designing to constructing and operating should be taken into account for large scale experimental facilities. To take high energy physics experimental center as an example, the system of accelerators should be integrated optimally. All the experimental devices and components, such as ion source, power source, magnet, controlling system, vacuum system and so on, not only need to work stably, reliably and efficiently for a long time, but also need good interfaces. Each kind of experimental facilities should work coordinately. The system of detectors should also be optimized, including the integration of software and hardware. Various experimental devices and components in the detection system, such as detecting devices, analyzing magnetics, electronics, computers and shielding systems, all need to work reliably. They should also cooperate with one another to ensure the stable operations of experimental apparatus.

From the perspective of the organization and management of large scale experimental facilities, the project needs a large number of working staff and experimental teams. They should cooperate with one another to accomplish the tasks. Thus team integration is also an important part of engineering integration.

10.3 Integration of Medical Imaging Techniques

Medical science should prevent, diagnose and treat human disease. Based on correct diagnosis of disease, the disease can be effectively treated. The common diagnosis method includes the analysis of physiological samples and the measurement of physiological indicators. The development of science and technology makes medical diagnosis looking through the human body without trauma, observing and photographing human's inner structures and functions. Medical imaging techniques has already become an important way of contemporary medical diagnosis.

Radiation imaging exploits the interaction between rays and human body to detect human inner structures and functions. On the aspect of structural imaging, we can use rays to irradiate the human body. Rays can penetrate the body and the penetrated rays can be measured. The rays have different degrees of penetration for different parts, which provides us with the information about inner structure of the body. On the aspect of functional imaging, tracers with low radiation can be injected into the human body. They participate in human activities, which can be detected from outside the body. Tracers of different kinds participate in different activities in different parts of the body, which can provide us with the information of a human's inner function. We can scan the human body layer by layer from outside the body to get layered images, and then rebuild the stereoscopic images. These human inner structural and functional images provide us with direct evidence for medical diagnosis.

Presently, we have already had many kinds of medical imaging techniques. The commonly used medical imaging techniques include CT, MRI, fMRI, PET, SPECT, etc. (Tang, 1999; 2001).

On the aspect of structural imaging, CT and MRI can examine the internal structure of human body without causing damage. The former uses X-rays or γ-rays to irradiate the human body, gaining the picture of internal structure of human body through multiple projections. The latter exploits the principle of nuclear magnetic resonance to examine the distributions of protons in the human body to gain the images of internal structures. These imaging techniques are vital for diagnosis and treatment of diseases. However, these images are structural images instead of functional images.

Functional image is different from structural image. It can show the regions of the human body where the function has changed and their temporal and spatial features. If we want to know the real-time brain functional activities and the functional connections of these brain regions, or if we want to know the change of brain functions when we are suffering from brain disease, we should adopt brain functional imaging. In recent years, based on MRI, functional MRI and NMR spectroscopy techniques are frequently used as important tools of functional imaging.

The principle of brain functional magnetic resonance imaging (fMRI) is that the nuclear magnetic resonance signal is related to the oxygen level in the blood. To examine the change of oxygen level of blood in the brain, we can know the change of activities of neural cells, because when the individual conducts cognitive missions, the brain region will be excited, with the blood flow increasing, while the oxygen reduction is less than the increase of blood flow, the deoxygenated hemoglobin decreases. Deoxygenated hemoglobin is a paramagnetic substance, which can extend the characteristic quantity T2, and then the T2 weighted imaging signal will increase.

(Nuclear magnetic resonance signal has many characteristic quantities, such as T1 and T2 as relaxation time; T1 is vertical relaxation time while T2 is horizontal). This effect is called the blood oxygen level dependent effect, which is the basis of brain functional imaging using nuclear magnetic resonance (Logothetis et al., 2001). Additionally, the principle of NMR spectroscopy is that, the spectrum of various chemical molecules in relevant regions can be measured by examining the chemical displacement of nuclear magnetic resonance signal.

Chapter 4 has already mentioned nuclear medicine. SPECT and PET are important techniques of brain functional imaging. The former is to inject the chemical labeled with nuclide emitting γ-rays into human body, and to measure γ-rays from outside of the body to gain the layered images of the distribution of this kind of labeled chemical. The latter is to inject the chemical labeled with nuclide emitting positrons into the human body, and to measure γ-rays emitted after the annihilation of positrons in order to gain the layered images of the distribution of this kind of labeled chemical. Usually, we use glucose labeled by 18F to get the metabolism image inside the brain, which can quantitatively measure the metabolism function of glucose; or use water labeled by 15O to gain the blood image in the brain, which can identify the brain activation areas of cognition activities. These techniques can be used to gain the three-dimensional images of brain functions with the spatial resolution about several millimeters.

The following are the features of PET: three-dimensions (which can increase sensitivity and accuracy), living body (damage-free, natural physiological state), dynamics and functions, quantitation, short lifetime positron nuclide as a part of living organism, diversity and specialty of tracer materials. However, the brain functional images gained in this way has low spatial resolution, which needs to be combined with structural images of high resolution. This technique can be used to detect the changes of local energy consumption, and the features of the distributions of neural activities in different brain regions, so the brain regions where the functional activities happen, such as the brain regions of visual sensation, auditory sensation, language, and thinking, can be known, and the brain regions of schizophrenia and aphasia can also be located.

On the aspect of brain functional imaging, besides the former techniques, there are other imaging techniques, such as multiple optical imaging techniques based on the change of optical property inside the brain, which can provide us with high resolution images for observing functional construction in the cortex. Near infrared spectroscopy and optical coherence tomography develop quickly and are very promising. The former exploits the effect of brain activities on the transmission of near infrared light to form images, while the latter uses the principle of optical coherence to conduct tomography of brain organizations.

The images gained through medical imaging technology include human structural images, and all kinds of human functional images. Human brain functional images include brain functional images of resting state and those of conducting various tasks. The features of medical images include spatial and temporal resolution. The imaging device should give pictures of high resolution which are important to accurately analyze the gained images to better diagnose and treat diseases.

There are integration phenomena in each level and on every aspect of the medical imaging technique. Every kind of medical imaging technique is the integration of hardware and software. On the aspect of hardware, we can find the integration of radiation sources, detectors, electronics, and computers. Each component should fit with other components in properties. The properties of components will influence the image resolution. To improve the image resolution, the bottleneck problems should be solved first. On the aspect of software, we can find the integration of image segmentation, match and fusion. From the perspective of operations of medical imaging facilities, there is the integration of data acquisition and data processing. On the types of medical images, we can find the integration of structural images, physiological function images, and functional images relevant with tasks.

Different kinds of medical imaging techniques are called different modalities. For example, CT and MRI are two kinds of modalities. The diagnosis data of different modalities can complement each other. The developing trend of medical imaging technology is the integration of multiple modalities. It is reasonable and efficient to diagnose and treat diseases coordinately using multiple imaging techniques. One example of this aspect is the construction and operation of PET/CT device. This kind of device integrates the PET imaging technique with the CT imaging technique, integrating structural imaging with functional imaging. Another example is construction and operation of the radiation treatment device combined with the ray imaging technique. This kind of device combines medical imaging the technique with the radiation treatment technique, integrating medical diagnosis with medical treatment.

Another developing trend of medical imaging technique is imaging of activities at the molecular level within the human body, which is called the molecular imaging technique. This technique can be integrated with the traditional structural imaging technique and the functional imaging technique. The human internal image can be gained both at the macroscopic level and at the molecular level, which helps understand the molecular mechanism of the disease, better diagnose, and treat it. The factors of the molecular imaging technique include molecular probe, amplification of signal, high sensitive detection and analyzing software. These aspects need to be efficiently integrated.

Chapter 11　Education Integratics

There are many integration phenomena in the field of education. Applying general integration theory to this field and studying the features of integration phenomena in the field of education is very practical.

We propose a new discipline termed as education integratics which studies the integration phenomena, principles and applications in the field of education. Meanwhile, we want to construct another new discipline termed as intelligence integratics which studies the integration phenomena, principles and applications of human intellectual activities. This chapter will discuss the questions of education integratics, i.e. the integration of intelligence and the integration of contents on education.

11.1 Integrated Intelligence

Intelligence is the integration of human mind abilities and behavioral abilities. The development of intelligence is the process of integration of all kinds of abilities. This process of integration has its own rules, based on which the education can be conducted.

In the book *On Intelligence — Integration of Mental and Behavioral Abilities* (Tang, 2010), the integration of intelligence has been profoundly discussed. The following is the abstract of the relevant parts of that book:

> Intelligence is a very complex phenomenon. There are various kinds of integration phenomena in intelligent activities. Thus it is necessary to discuss the integration phenomena of intelligence, focusing on the integrative actions and integration processes. The integrative actions of intelligence are important mechanisms of intelligence activities, and the integration processes are important contents of intelligent activities.
>
> Intelligence integratics exploits the structure and process of intelligent activities with the view of integration, especially the integrative action and integration processes to explore the nature of intelligence. As a theory, intelligence integratics is a kind of theory about the nature of intelligence, mental abilities and behavioral abilities. As a discipline, intelligence integratics studies all types and levels of integrative phenomena, rules and applications in intelligent activities.
>
> The range of research on intelligence integratics is the phenomena of intelligence, mainly involving the integration phenomena related to intelligence. Next, the features of intelligence integratics will be illustrated on the aspects of research objects, concepts, contents and approaches.
>
> From the perspective of research objects is concerned, the research objects of intelligence integratics are different from those of previous intelligence theories. Intelligence includes mental abilities and behavioral abilities. In intelligent activities, there are many mental interactions, such as the interaction of psychological

components, mind-brain interaction, mind-body interaction, mind-matter interaction, and mind-society interaction. The previous theories on intelligence only paid attention to several abilities, investigated a few components of intelligence, involved several kinds of mental interactions of mental activities, while intelligence integratics studies all the components of intelligence and the integrations of all kinds of mental interactions.

From the perspective of theoretical concepts, they are different from previous theories on intelligence. The core concept of intelligence integratics is integration. The intelligence phenomena involve components of intelligence, integrative actions, integration environment, integration process and various intelligence integrated body. Different kinds of components of intelligence undergo different forms of integration processes through their mutual interactions, constructing intelligence integrated body of different levels and properties, emerging new features under some conditions. Previously, some theories on intelligence, such as the factor analysis theory, analyzed the factors of intelligence, but did not discuss their correlations and development, while intelligence integratics stresses the integration of all levels of the components of intelligence involved in intelligent activities and the integration of all kinds of mental interactions in intelligence activities.

On the aspect of research contents, intelligence integratics is also different from previous theories on intelligence. It believes that intelligent activity is not the simple addition of components of intelligence, but integrated by their interactions. Thus the research content emphasizes the integrative action and the integration process of intelligence. There are various kinds of integration phenomena in the intelligence activities. Due to the diversity and complexity of components of intelligence, integrative actions, integration environments and integration processes, the intelligence of different people are totally different.

Based on the inborn genes, the individual intelligence develops through all kinds of mental interactions in the process of long-term integration practice. The experiences show that completed intelligent components, active integrative actions, enriched integration environment and coordinating integration process are the keys to effectively developing intelligence.

As far as the research approach is concerned, the research approach of intelligence integratics is also different. The previous research approach and the concrete intelligence theory usually emphasizes one side or several sides of intelligence while intelligence integratics holds that intelligence is an integrated whole of multiple components of intelligence and that intelligence activities have various integrations of mental interactions. Thus we should integrate various intelligence research approaches and concrete theories on intelligence to construct correct theories. Intelligence integratics has integrated all kinds of research approaches and concrete theories on intelligence.

There are two aspects of the research on intelligence integratics. One is the integration phenomena of intelligence, especially the integrative actions and integration processes in intelligent activities, and another is the integration of theories on intelligence which include the integrations of contemporary intelligence research approaches and various concrete theories on intelligence. The aim of studying the integration of intelligence and the theories on intelligence is to understand the nature of intelligence and to explore the methods of raising the level of intelligence. Thus the study of intelligence integration and study of theories on intelligence are consistent.

On Intelligence—Integration of Mental and Behavioral Abilities (Tang, 2010) proposed a theoretical framework of intelligence, including the definition of intelligence in the general sense which is based on mental interactions and unification theories, the unified research approach of intelligence, the views of intelligence structure and process, and intelligence integratics.

Intelligence integratics is one part of this theoretical framework of intelligence, investigating the intelligence phenomena and the nature of intelligence from the view of integration. Intelligence integratics includes the study of intelligence integration and the integration of theories on intelligence.

When it comes to the aspect of integration of intelligence, we would like to list the main points of intelligence integratics as follows.

Firstly, the basis of individual intelligence activities is brain and body. Intelligence activities include mental activities, behavioral activities and their couplings. Mental activity is the function of the brain, which controls behaviors fulfilled by the body. Individuals interact with external environment through sensational organs, locomotive organs, and language organs.

Secondly, intelligence activities cannot be separated from the environment. There are various kinds of interactions in the unification of mind-brain-body-natural environment-social environment. Individual intelligence activities proceed in the natural and social environment. The environment will play an important part in intelligence integration.

Thirdly, intelligence is the integration of mental abilities and behavioral abilities as they are closely related. Their combination constructs intelligence. Mental ability and behavioral ability are both integrated through integration processes of different levels and types. They have complex structures.

Fourthly, the components of intelligence and concrete abilities are multilayered and diverse.

Mental ability is one aspect of intelligence, integrated by arousal-attention ability, cognition ability, emotion ability, volition ability, etc. Among them, each component of intelligence further includes many specific abilities. For example, cognition ability includes sensory ability, perceptual ability, memorizing ability, thinking ability and language ability, etc. Each specific ability has its structure. For example, thinking ability is composed of analyzing ability, synthesizing ability, comprehension ability, reasoning ability, etc.

Behavioral ability is another aspect of intelligence, integrated by locomotive ability, operational ability, adaptive ability, social ability, etc. Each component of intelligence includes many specific abilities, for example, social ability includes interpersonal ability, managing ability, expressing ability, etc.

Intelligence is not a single component or a single ability. All kinds of components of intelligence and specific abilities combine with each other. Intelligence is a complex unification composed of various components of intelligence and specific abilities. We cannot describe complex intelligence with only one index.

Fifthly, intelligence activities have various kinds and levels of integration and are realized by many kinds of mental interactions.

On the aspect of mental activities, arousal-attention, cognition, emotion, volition interact with one another and integrate into intelligence activities through mind-brain interactions. On the aspect of cognitive processes of mental activities, sensation, perception, memory, thinking and language integrate into cognitive activities through their interactions and mind-brain interaction. On the aspect of thinking processes of cognitive activities, analyzing, synthesizing, understanding, and reasoning integrate into thinking activities through mutual interactions. From the perspective of behavioral activities, there are similar situations. In intelligence activities, all kinds of mental interactions integrate different components of intelligence, and these integrations are not random additions but organic unification.

Sixthly, there are integration processes of different levels and types in mental activities and behavioral activities. During the integration process of intelligence activities, many different components of intelligence form various kinds of unifications at different levels. Mental activities and behavioral activities both have complex processes. There are usually the phenomena of optimization, synchronization and coordination in the integration processes of mental and behavioral activities.

Seventhly, the integration process of intelligence is active. Mind and behavior form a unification through active integration. For example, long-term memory is not the simple piling up of events but an active process of integration, as we know that there are networks with classifications and connections through active integration processes of organizing materials in memory. Under some conditions, during the integration process of intelligence, there will be new functions. For example, in the long-term active thinking activities, there might be new ideas, which can lead to new results.

Eighthly, the development of intelligence. According to the theory of evolution, human beings' intelligence is the product of evolution. Individuals develop and grow with their components of intelligence and specific abilities developing constantly through integration processes of intelligence during practice. Thus, the overall intelligence which integrates the essence of various kinds of components of intelligence will constantly develop. Intelligence integratics is an ongoing process. It is a staged process. We should study the rules of development of all kinds of components of intelligence and concrete abilities.

Ninthly, there are individual differences in intelligence, because of the different inborn genes, different postnatal practices and the diverse components of intelligence, integrative actions, integration environments and integration processes. As different individuals have diverse components of intelligence and specific abilities, they have both advantages and disadvantages. We cannot require different individuals with the same criteria.

Tenthly, the cultivation of intelligence. Individual intelligence can be cultivated and intelligence level can be raised. Many factors can influence intelligence. Besides heredity, nutrition, environment and education, conscious learning and practice are decisive towards the development of intelligence. The development of intelligence does not depend on one single factor but on many aspects. It is necessary to learn and practice actively in the long run to promote coordinating development of many kinds of components of intelligence and concrete abilities.

On the integration of intelligence theories, intelligence integratics includes the following main points.

The first one is the development of intelligence theory. Intelligence integratics stresses the integration process of theories on intelligence, and believes that the development of intelligence theory is based on existing knowledge about intelligence and that it is an integration process of new phenomena, concepts and theories. As people expand their knowledge about intelligence, the theory of intelligence constantly develops.

The second one is the integration of intelligence research approaches. Contemporary intelligence research has many research approaches, which emphasize different kinds of mental interactions in intelligence activities. The unified research approach of intelligence based on mental interactions is a new research approach integrating all kinds of research approaches in contemporary intelligence research.

Intelligence integratics is a kind of intelligence theory based on this research approach. It studies various mental interactions comprehensively and stresses their unification.

The third one is the integration of specific theories on intelligence. Intelligence integratics is the theory of integration of all the fruits of concrete theories on intelligence. Many contemporary theories on intelligence have their basis and advantages, and they can complement with one another. Intelligence integratics integrates the fruits of these specific theories on intelligence, including the cognition theory, factor analysis theory, biology theory, emotion theory, embodiment theory, situation theory and society theory of intelligence.

Although intelligence integratics includes large amounts of relevant contents about concrete theories on intelligence, it is not the simple collection of them. Instead, it integrates the fruits of these concrete theories on intelligence and constructs new theories based on mental interactions and their unifications.

These earlier discussions about the features of intelligence integratics are enlightening on education. Education should accord with these features to effectively promote the integration of mental abilities and behavioral abilities.

11.2 Integration of Contents of Education

Human quality includes morality, intelligence, physique and aesthetic. The previous section has discussed intelligence, which is one part of human quality. The aim of education is to shape all-round developed people. The well-rounded education should integrate these aspects.

Morality education is to develop human's morality, that is, everyone should love his motherland and his people, have lofty ideals and noble moralities, respect the old and care about the youth, and obey the rules and laws. Intelligence education is to increase humans' knowledge and abilities. Everyone should love science and learning, have the right way of thinking, acquire scientific knowledge and skills, cultivate the ability of solving practical problems, and serve the society. Physique education is to cultivate people's strong body. Everyone should do exercises and develop healthy mind and body. Aesthetic education is to teach people understanding the standards of truth, goodness and beauty, and having the strong will to pursue good ideals.

People's morality, intelligence, physique and aesthetic are connected and unified. They should be developed coordinately. The integration of education stresses the integration of these qualities. People will be shaped comprehensively through the integrated well-rounded education. People are different with their own traits. Thus, education needs to cater for these differences, and cultivate people with their own characteristics. However, the comprehensive development of morality, intelligence, physique and aesthetic is necessary for everyone, whose strengths should be promoted based on comprehensive development.

Educational neuroscience (or neuro education) is the cross-discipline of neuroscience and education science. This subject conducts neuroscience studies of educational theories and educational practice. Morality, intelligence, physique and aesthetic all have relevant neural basis. The function of the brain has uniformity, so does the neural basis of the human quality. To study the mechanism of brain activities of educators and students will make the research content of neuroscience more diverse.

Educational neuroscience also develops educational theory based on the research achievement of neuroscience and implements educational theory into educational practice. The knowledge about the neural basis of the human quality and the uniformity of brain function is the theoretical basis of integration of contents of education. Applying the research results of brain science into educational practice will promote human quality more effectively.

In the cross-disciplinary studies of neuroscience and educational science, other subjects are also involved, such as physiology, psychology, medical science, cognitive science, managing science, etc. Large amounts of knowledge integration of many subjects are included.

The development of the brain has certain sensitive periods. Educating children in these particular periods will have the best effect. The periods from infant, child to adolescent are the key stages for cultivating the human quality. The well-rounded education should begin with the infants, making them develop healthily from the very beginning. Schools of every level should create an environment in which the development of morality, intelligence, physique and aesthetic should be emphasized, and conduct comprehensive well-rounded education.

However, education does not confine itself to children and adolescents. Middle-aged and old people also need to constantly develop their quality. The well-rounded education should last for all one's life. Society also needs to conduct well-rounded education towards middle-aged and old people. Old people should always learn and contribute to the society. The well-rounded education is not only the mission of schools, but also the task of the whole society. Social environment is the broad environment of shaping people. It is necessary to create an atmosphere which benefits learning and comprehensive well-rounded quality development.

How to integrate the education contents of the human quality in education practice is a topic which awaits further research. In recent years, some educationalists recommend the educational method of "learning by doing", i.e. a kind of scientific education during real practice, and they have gained excellent results (Wei & Rowell, 2005). From the perspective of comprehensively cultivating the quality of morality, intelligence, physique and aesthetic, the content of "learning by doing" can be expanded to scientific knowledge, and the development of morality, physique and aesthetic as well.

Appendix

Some Parts of Sherrington's Descriptions About the Integrative Action of Nervous Systems

In his book *The Integrative Action of Nervous Systems* in 1906, Sherrington emphasized the importance of integrative actions of nervous systems.

He suggested adopting three viewpoints to study neurophysiology.

The first one is neural nutrition. Living neural cells undergo physiological activities similarly with other living cells. The nutrition of living neural cells and nervous systems should be investigated.

The second one is neural conduction. Neural cells can transmit nervous impulses. The transmitting process should be examined.

The third one is the integrative action. The neural system in a multicellular organism which connects together with various organs in the body.

Among these three viewpoints, an integrative action is especially important.

He believed that there exist many kinds of integration in a living body. First and foremost, structurally speaking, there are integrations of animal organs and also some integrated animals by single cells.

The second kind of integration is chemical integrations, for instance, the coordinating activities of various kinds of glands. The third kind of integration is the integrative actions of blood circulation, which helps achieve the unified activities of the body. The fourth kind of integration is the integrative action of nervous systems.

He pointed out that the integrative action of nervous systems are different from the other kinds of integration in that they are not achieved by the transmission of materials between cells, but through the conduction of neural signals, so they can transmit at high speed for a long distance. Neural connections have fine spatial distributions, and thus neural conduction has a precise temporal distribution.

He conducted systematic researches on the activities of central nervous systems. In his book *The Integrative Action of Nervous Systems*, he proposed the famous conclusion that integrative actions are vital for the functions of central nervous systems. He has studied the integrative action of nervous systems at different levels.

Beginning from the spinal reflex, through the reflex of tensional muscles and flexion reflex, he conducted in-depth studies about the integrative actions of central nervous systems. He believed that reflex is one fundamental mode of activity in the central nervous system. The sensor receives external stimulus and turns it into nervous impulses. Then the impulses transmit to the central nervous system through afferent nerves, and in turn the central nervous system transmits the impulses to the peripheral effectors and finally the organs move. This is called reflex.

Thus reflex includes input, output and central nervous systems. Organisms have three kinds of structures, namely sensors, central nervous systems and effectors. They correspond respectively to three processes, i.e. the reception of external stimulus, the transmission through central nervous systems, and the output process. These three processes comprise reflex activities.

Reflex activity is a unit of neural integration. Each reflex is an integrated reaction. The neural activities without reflex are not complete integration processes. In a simple reflex, such as knee jerk reflex, there is integrative action. He used the concept of coordination to talk about reflex. Coordination is important for a reflex process and a reflex is the result of coordinating each kind of inputs. He also pointed out that nervous systems have the excitatory process and the inhibitory process. The response of central nervous system is integrated.

He discussed all kinds of features of spinal reflex in detail, including the coordination of simple reflex, the interactions between reflexes, the simultaneous combination in compound reflexes, the successive combination in compound reflexes, the adaptive reactions of reflex, etc.

He also discovered the interactive neural control while studying reflexes. When one muscle contracts, another related one relaxes. From the perspective of neural activities, when the motor neuron which controls the contraction of muscle is active, another motor neuron which controls the relaxation of muscle is inhibited. The excitable activities are accompanied by inhibitive activities, which is called interactive neural control.

At the level of synapse, he discussed the integrative actions of synapses, and first proposed the concept of synapse, which is defined as the contact point between the end of one neuron and the dendrite or cell body of another neuron. It is the basis of neural signal transmission. Synapse transforms many inputs into one output, which is the integrative action of synapse. He pointed out that there are interactions of excitable and inhibitive functions in the synapse. Each synapse is a coordinating mechanism.

At the level of neuron, he discussed the integrative action of single neurons. The integrative action of one motor neuron is manifested as the integration of signals by cells. One motor neuron can integrate excitatory input and inhibitory input, appraise all kinds of signals and decide the behaviors. Thus he treated single neurons as the cellular basis of integration. The overall integrative action of the brain can be studied from a single motor neuron.

At the level of the integrated organisms, he pointed out that the individual multicellular animals are not only the summation of many organs. These organs are unified to be consistent through integrative actions of nervous systems.

He not only discussed the coordinating activities of neurons at the level of central nervous reflex, but also applied these principles to upper level processes. For example, on the aspect of sensation process, he discussed the phenomena of vision.

He studied the integrative actions of the brain cortex, pointing out that the brain is an integrated whole. He also discussed the integration of mentality and body, believing that the highest level of integration in the nervous system is the integration of mentality and body.

On the aspect of research methods, he integrated different kinds of methods, combining neuroanatomy, neurophysiology and behavioral studies to conduct researches on nervous systems.

The Origin of General Integration Theory

The concept of general integration theory is formed gradually in the process of my research and teaching.

In the 1980s, I was encouraged by elder scientists to start the cross-disciplinary studies of physics, biology and medicine. Later I began to show interest in the cross-disciplinary studies of physics, brain science and cognition. In this period, I got to know many friends in these fields and discussed intensively with them, and got enthusiastic assistance from them on experiments and theories.

In the 1990s, due to the need of national scientific research, my collaborators and I participated in the National Climb Plan — "The Basic Studies of Advanced Technologies in Nuclear Medicine and Radiotherapy", and the major program of Natural Science Foundation of China — "Developing the Near-field Technique and Studying the Features of Biological Macromolecule System".

Both nuclear medicine and radiotherapy apply nuclear techniques into clinical diagnosis and treatment to serve people's health. The program of "The Basic Studies of Advanced Technologies in Nuclear Medicine and Radiotherapy" included nuclear medicine, radioactive medicine and radiotherapy. This program involved nuclear physics, radiochemistry, medicine, pharmacology, computer science and so on, with researchers of many subjects participating in it.

The program of "Developing the Near-field Technique and

Studying the Features of Biological Macromolecule System" studied techniques of nanometer detection and manipulation and applied them into the studies of systematic features of biological macromolecule system. This program involved physics, molecular biology, chemistry, nano technique, and accurate machine technology. It also involved many researchers from different fields.

I gained knowledge and techniques of medical physics and biological physics by participating in these two programs. Meanwhile, I came to form the concepts of knowledge integration and technology integration, and recognized the importance of integration in cross disciplinary studies through the cross-studies of biology, medicine, physics and so on.

Cross-disciplinary studies are not just packing together the previous topics and equipment of several different disciplines. In fact, cross-disciplinary studies require the long-term collaborations of researchers from different subjects to solve the same scientific and technological issue. The process of cross-disciplinary studies is an integrative action, which calls for knowledge integration and technology integration of different fields and also resource integration and management integration in teamwork.

At that time, due to the necessity of applying medical imaging technique into the research of brain function, I began to learn brain science and cognitive science. The human brain is the most complex system in nature. The studies of brain science include detection, understanding, protection, exploitation, and mimicking of the brain. Each field of brain science involves many subjects. Brain functional imaging technique is to measure the activities and connections in the brain regions in the resting state and the task state, and to study cognitive processes without traumas to know the working mechanism of the brain. To conduct experiments of brain imaging, it is not enough to have knowledge or an experimental technique of one subject, there must be integration of knowledge and techniques of many disciplines.

In the process of learning brain science, Sherrington's book *The Integrative Action of Nervous Systems* exerted great influence on me. He said that the main function of central nervous systems is integrative action. The central nervous system of an animal unifies separate organs into a consistent individual animal. Through spinal reflex processes, he conducted in-depth analysis of the integrative actions of the central nervous system. He also discussed the integrative actions of single nervous cells and synapses.

His thoughts persuaded me to try and investigate various activities at different levels inside the brain with the concept of integrative actions. The facts show that from molecule, gene, synapse, neuron, neural circuit, brain region with specific function, functional subsystem to the whole brain, there are different kinds of integration phenomena at different levels of the brain.

At the system level, the brain has four functional systems, namely, one functional system to maintain arousal, one to process information, one to regulate and control, and one to appraise and feel. The coordination of these systems ensures the normal activities of the brain. Besides the structural and functional integrations of the brain, there are also information integration and psychological integration inside the brain.

At that time, besides conducting experiments of brain imaging, my colleagues and I also worked on neuroinformatics, which is the cross-discipline of neuroscience and information science. One aspect of it is to study the carrier form of neural information and the generation, transportation, processing, encoding, storing and retrieval of information in the brain from the view of information and information processing.

Another aspect of neuroinformatics is to use modern information tools to gather the research data at different levels of the brain and to build the database and working station of neural information, analyze the data, construct models, and share scientific data to cooperate with others. Scientific data sharing in many disciplines, including the sharing of neural scientific data, is the integration of scientific information and Internet communication.

At the beginning of 2001, I worked full-time at Zhejiang University, and organized the brain and intelligence research center. At that time, because of the requirement for discipline construction, my colleagues and I organized and conducted such programs as, "Research and Application of Brain and Cognition" and "Research on Language and Cognition" at Zhejiang University.

The program of "Research and Application of Brain and Cognition" conducted interdisciplinary studies concerning several subjects and applications in brain and cognitive science, including molecular neurobiology, neural-endocrine-immune network, neuroinformatics, and the applications of cognitive science. The program of "Study on Language and Cognition" conducted cross-disciplinary studies of many frontier issues about language and cognition, including mentality and awareness, language cognition, social cognition, and cognitive science.

Afterwards, Zhejiang University established the Center for the Study on Language and Cognition (CSLC), which is the innovative base of philosophy and social science. Working in this center gave me a chance to learn from experts in humanities and social science. We have discussed widely to promote the combination of arts and science, granting me some experiences about the integrative research of natural science, humanities and social science.

In the aforementioned two programs, I conducted some specific projects. From the existing experimental facts, I explored the essence of mentality, the theoretical system of psychology, and the integration of cognitive science theories, which made me think about integration phenomena in many other fields.

Mind is a very complex phenomenon, including components of arousal, cognition, feeling, will, etc. I felt that it was necessary to study the integration of these components and their mutual interactions, and the integration of mind, brain, body, natural environment and social environment to comprehensively understand the nature of mentality.

Modern psychology includes many basic disciplines and applied disciplines with many branches. The basic disciplines include experimental psychology, biological psychology, physiological psychology, neural psychology, psychological physics, cognitive psychology, developing psychology, personality psychology, social psychology, etc., while the applied disciplines include educational psychology, school psychology, medical psychology, clinical psychology, engineering psychology, industrial and organizational psychology, sports psychology, military psychology, advertisement psychology, judiciary psychology, etc. I have proposed the grand unification theory of psychology based on psychological interactions, trying to integrate the enormous disciplines into a unified theoretical system.

Contemporary cognitive science relies on many research approaches, mainly including the neurobiological approach, information processing approach, embodied cognition approach, situated cognition approach, social cognition approach, evolutionary psychology approach, developing psychology approach, artificial intelligence approach, etc. I believe that the research of cognitive science should integrate all these research approaches to construct the integrative theory of cognitive science.

For the last ten years, I have collaborated with several colleagues and supervised a number of graduate students to conduct experiments on the brain activities in the resting state and the cognitive state of the brain. We have especially cooperated with the key laboratory of medical physics and engineering of Peking University. The work of this laboratory in recent years includes the development and application of CT, MRI, SPECT, PET, fMRI, on the aspect of hardware, and image segmentation, registration, fusion on the aspect of software. It also includes the research on radiotherapy.

We have performed structural imaging and functional imaging of the brain, and promoted the research of molecular imaging, which is the cross-discipline between medical imaging and molecular biology, using various medical imaging techniques to form images of specific molecules inside the human body without trauma.

These experiences have enhanced my understanding of integrative phenomena. I feel that technological integration can promote the development of new technology. In the field of medical imaging technology, this kind of integration is demonstrated by multimodal imaging techniques, multiple experimental data of images and multiple analyzing methods of images.

For the development of pedagogy, I have studied the issue of intelligence at Zhejiang University and discussed with teachers from Zhejiang University and Zhejiang Normal University about the educational neuroscience, and thus I gradually touched the concepts of intelligence integratics and educational integration.

Intelligence includes mental abilities, behavioral abilities and all kinds of specific abilities. Intelligence has hierarchical structures and dynamic developing processes. I believe that intelligence is the integration of mental abilities and behavioral abilities and also the integration of all kinds of psychological interactions, and that it is necessary to study various integrative actions at different levels of intelligence and the integration processes in dynamic intelligence activities.

Educational neuroscience is the cross-discipline between education science and neuroscience. In the field of educational neuroscience, on the one hand, we should conduct neuroscience research in regards of educational theories and practices; on the other hand, we should develop educational theories and practices based on the research fruits of brain science. Educational neuroscience needs the integration of both neuroscience and educational science.

In the aforementioned practice, I contacted many integrative phenomena in different fields, and thus formed the concept of general integration theory. I have noticed that there are various integration phenomena in the fields of nature, engineering and technology, and human society. The commonality of these phenomena made me investigate further the features of all kinds of integrative actions, integration processes, and their general characteristics.

On the basis of learning from the brain and researching on the experimental facts of different fields, I have summarized some general concepts in integrative actions and integration processes, such as optimization, globalization, complementarity, coordination, synchronization, binding, emergence, adaptation, and grand unification to describe the common features of integration phenomena.

I believe that it is necessary to summarize the facts and concepts of integration phenomena in different fields and build a discipline which studies the general features and principles of various integration phenomena and their applications. I call this discipline as general integration theory or general integratics.

I have also considered applying general integration theory or general integratics into several specific fields to build a series of sub-disciplines studying integration phenomena in specific fields, especially the specific rules and their applications of integrative actions and integration processes. These sub-disciplines fall into special integratics.

These special integratics form a group of disciplines. General integration theory guides the integration of these sub-disciplines. They comprise many types. Some examples are given as follows.

In the field of biology, there should be a sub-discipline which studies the features, principles and applications of integration phenomena in biology, which can be called bio-integratics.

In the field of medicine, there should be a sub-discipline which studies the features, principles and applications of integration phenomena in medicine, which can be called med-integratics.

In the field of psychology, there should be a sub-discipline which studies the features, principles and applications of integration phenomena in psychology, which is called psycho-integratics.

In the field of cognition, there should be a sub-discipline which studies the features, principles and applications of integration phenomena in cognition, which is called cogno-integratics.

In the field of information science, there should be a sub-discipline which studies the features, principles and applications of integration phenomena in information science, which is called info-integratics.

In the field of geoscience or earth science, there should be a sub-discipline which studies the features, principles and applications of integration phenomena in geoscience, which is called geo-integratics.

In the field of space science, there should be a sub-discipline which studies the features, principles and applications of integration phenomena in space science, which is called space integratics.

In the field of environment, there should be a sub-discipline which studies the features, principles and applications of integration phenomena in environment, which is called environment integratics.

In the field of engineering, there should be a sub-discipline which studies the features, principles and applications of integration phenomena in engineering, which is called engineering integratics.

In the field of technology, there should be a sub-discipline which studies the features, principles and applications of integration phenomena in technology, which is called technology integratics.

In the field of education, there should be a sub-discipline which studies the features, principles and applications of integration phenomena in education, which is called education integratics.

In the field of economy, there should be a sub-discipline which studies the features, principles and applications of integration phenomena in economy, which is called economics integratics.

In the field of culture, there should be a sub-discipline which studies the features, principles and applications of integration phenomena in culture, which is called culture integratics.

In the field of society, there should be a sub-discipline which studies the features, principles and applications of integration phenomena in society, which is called social integratics.

Meanwhile, we should also establish many specific integratics in specific fields, such as neuro-integratics, brain integratics, knowledge integratics, intelligence integratics, management integratics, and so on.

To carry out studies of general integration theory and special integratics needs joint and long-term effort of experts from different fields. Therefore, I wrote this book to illustrate the theories and applications of general integration theory and asked suggestions from experts. I hope we can discuss deeply to form a relatively complete system of theory.

As mentioned before, the ideas of general integration theory are gradually formed under the influence of Sherrington's book *The Integrative Action of Nervous Systems.* In the light of his opinion, this book begins with the analysis of integration phenomena at different levels of the brain, and then investigates integration phenomena in nature, technology and human society from the proposal of brain integratics, neuro-integratics and brainics to the emergence of general integration theory.

During the process of writing this book, I read works of cybernetics, general system theory, information theory and theory of open complex giant system, and benefited a lot from the predecessors. I analyzed the relationship between the general integration theory and those theories proposed by the predecessors and tried to distinguish them. I hope the studies of general integration theory and a series of special integratics can complement and enrich the predecessors' theoretical achievements.

The Principles and Methods of General Integration Theory

Integration Phenomena and General Integration Theory

Various integration phenomena exist in nature, in human society and in the realm of thinking.

What is integration? Components within complex things are integrated into the unity of coordinating activities through internal and external integrations. Hierarchical complex things vary in internal integration components, integration effects and integration environment, leading to entities with different contents and forms. Integration is a dynamic process where new characteristics emerge under certain conditions.

Integration phenomena can be observed everywhere at anytime in our daily life. For instance, we make some relevant words into one sentence when we are speaking, we put all kinds of objects we see into a unified scene on certain occasions, and we organize relevant things and ideas to write a letter to somebody. These integration processes involve both conscious and unconscious processing, which is an integration of all of these involved activities.

Complex things have multiple functions and processes of integration.

Our world includes the inanimate physical objects, the living biological objects, and the spiritual and thinking world of human beings. In the inanimate physical world, there are many kinds of integration, such as the integration of materials, energies, structures, functions, information and movement. In the living biological world, the integration includes structure, function, information, life activities, and biological evolution. As for the spiritual world of human beings, integration lies in perception, memories, thoughts, intelligence, emotion and consciousness. Integration is also a common phenomenon in economic affairs, political affairs, culture, education and other fields of human society. In the economic and political field, there are integration of economic activities and integration of political activities. In the field of culture, there are cultural integration phenomena. In the field of education, there is integration of multicultural education. In the field of science, technology and engineering, there is integration in a corresponding way.

Integration is ubiquitous. General integration theory provides us with a way to understand integrated phenomena. It is a discipline that studies the common characteristics, general laws and applications of various phenomena of integration in the natural world, human society, as well as mind and thinking.

Various specialized fields exhibit distinct integration properties in their components, functions, environment, process and the integrated entity. The special integratics is a discipline that applies general integration theory to a specific field and studies the characteristics, rules and applications of integration phenomena in the specific field.

There is a lot of research on self-organization which in a way is a spontaneous process of integration. Compared with the theory of self-organization, the general integration theory is characterized by the fact that it not only investigates spontaneous integration, but also emphasizes initiative control, including intervention, optimization and organization through the integration process. In this sense, the general integration theory pays more attention to integration in man-made systems.

Principles of General Integration Theory

From the characteristics of various integration phenomena, we try to summarize three laws for general integration theory—the law of pluralism and unity, the law of hierarchy and emergence, and the law of development and optimization.

(1) The Law of Pluralism and Unity in General Integration Theory

Complex things have complexity and diversity. Varied interactions, movements and changes of things constitute a rich and colorful world. Complex things, pluralistic yet integrated, show multiple structures and intrinsic holistic property.

The multiple characteristics of complex things refer to the integration aspect. They have many components, not just one, and each component has its own characteristics. Complex things are not simple superposition of different ingredients, but a holistic unity which is united into one through integrations. The united characteristics here refer to the existence and integrations of multiple components, which are complementary to one another and form the overall movements of complex things through harmonic activities.

Take consciousness as an example. Consciousness, a very complex phenomenon, is the integration of various factors in the human brain. From the perspective of neurobiology, the neural basis of consciousness is the neural network integrated with many neural subsystems. From the perspective of psychology, consciousness is a kind of psychological activities integrated with many psychological elements, and it is also the integration of mind, body and environment. From the perspective of informatics, consciousness is a complex entity of massive information integration, or in other words, the integration of conscious information processing and unconscious information processing. From the perspective of evolution, consciousness is the integrated optimization function through the long-term evolution of the brain. From the socio-cultural perspective, consciousness is the integration of human experience, culture and society.

(2) Law of Hierarchy and Emergence in General Integration Theory

Complex things have hierarchical a structure with multiple levels, rather than a single level. For integrated multiple entities, the complex levels are integrated by simple and lower levels. Different levels have different structures and movements. Each level has its own characteristics and rules interacting with other levels.

When the components of the lower level are integrated into the upper level, emergence will occur under the critical conditions. The emergence refers to the new characteristics that the original components do not have and the critical conditions are the conditions required for new characteristics.

Take the integration of life phenomena as an example. Life phenomenon covers multiple levels, including biological molecules, genes, subcellular structures, cells, tissues, organs, biological individuals, biological groups, and finally ecosystems. Different levels of integration and emergence appear at different levels of life phenomena. For example, various life activities emerge when biomolecules are integrated into living cells, involving material transport, energy metabolism, and information transmission in living cells and exhibiting new characteristics that do not exist at molecule levels.

(3) Law of Evolution and Optimization in General Integration Theory

Everything is constantly evolving. Integrated entities are formed and developed in the process of integration.

Things are constantly optimized during integration due to the existence of comparison, competition, selection, and reconstruction. That is to say, the integration process is also a process of optimization. The integration with optimization is an effective integration.

Take biological evolution as an example. Organisms evolve on the basis of heritage genes and long-term integration with the environment. Evolution plays a role of a tinker for organisms, which optimize themselves constantly in this process, as a result of natural selection.

Methods of General Integration Theory

General integration theory provides us with new perspectives and methods to study complex things. To be specific, based on this theory, we are able to observe things in a pluralistic and unified view while solving problems by reasonable reduction and systematic integration, to observe things with the knowledge of hierarchy and emergence while solving problems with corresponding methods, and to observe things in the view of evolution and optimization.

(1) Reasonable Reduction and Systematic Integration

Reasonable reduction and systematic integration are indispensable to complex things. Reduction starts from a complex whole and ends with its components, and integration needs observations of all the elements rather than one. We first investigate each component within a thing and then the complex network that all components make up.

The various components need to be analyzed in the reduction process at an appropriate, reasonable and meaningful level which is not unlimited. In this regard, it is called rational reduction.

Components will be integrated and such a process is not simple superposition but based on the understanding of internal interactions and relations between components as well as external interactions with the environment. It is called systematic integration which follows the things as they are.

Cells, for example, can be reduced to four subsystems, namely, nucleus, cytoplasm, cell membrane and cytoskeleton. The nucleus is the genetic material of the cell. The cytoplasm contains biomolecules, ions, water molecules, etc., which are integrated into various organelles and whole cells. The cell membrane consists of plasma membrane, nuclear membrane and many membrane structures. The main components of cytoskeleton are microfilaments, microtubules, intermediate fibers, and so on. These four subsystems interact with each other in a living cell. A living cell as an entity is a complex system which is integrated from these four subsystems. The coordinative activities of these subsystems guarantee the normal activities of a living cell.

(2) Hierarchy Analysis and Emergence Observation

We need to investigate the hierarchical phenomenon and emergence during the integration process in complex things. We need to know the different components and movements at different levels and the emergence of new characteristics under certain conditions.

The hierarchical analysis should reveal the hierarchical structure of complex things. We need to figure out characteristics of each level and their mutual relationships.

Emergence observation focuses on different emergence phenomena at each level, especially on critical conditions, new characteristics, and how the system gives rise to the new characteristics at different levels.

Take consciousness as an example. The brain and mind are very complex, with multiple levels and multiple dimensions. A good knowledge of human consciousness is used to figure out the characteristics and integrations of various components of the brain and mind at different levels, their interactions with body, environment and society, as well as the integration and emergence phenomenon. From the perspective of psychology, consciousness is the integration of factors like arousal, content, intention and emotion. From the perspective of neurobiology, consciousness is the integration of factors like arousal level, information processing, global broadcasting and attention enhancement. From the perspective of informatics, consciousness is the integration of body and information, the concept of which covers ego, environment, society and culture. To sum up, the integration theory of consciousness is the integration of psychological integration, neural integration, information integration and mind-body-environment integration.

(3) Dynamic Analysis and Optimization

Integration is a dynamic process which develops along with interactions among internal components and interactions with external environment. We should investigate the evolution process of complex things and make a dynamic analysis of the development during integration.

It is important to comprehend the characteristics of different stages during the course of evolution, and then to integrate things step by step. As mentioned earlier, general integration theory not only studies spontaneous integration phenomena, but also emphasizes the active integration process.

Integration is the process of optimization. When studying complex things, we should put forward various plans on the basis of a comprehensive understanding of the situation, and then evaluate and compare them repeatedly and finally choose the optimal solution which will be constantly adjusted in the implementation. Optimization in integration should be done in an active and planning way.

Take knowledge integration as an example. We need to constantly accumulate knowledge. One complicated scientific problem is often divided into several stages and carried out step by step. Knowledge integration is a process of exploration. In the long-time practice, knowledge is accumulated, experience is summed up and errors are corrected, so that new results can be obtained. We should not only organize existing knowledge, but also probe into new problems, create new concepts and opinions, put forward new assumptions and plans, and develop new knowledge through experiments. This is the innovation of knowledge.

References

Chinese Reference

Bao, H. (2003). The theory of biomedical knowledge integration(I). Medical Information, 16(6), 274-279. [包含飞. (2003). 生物医学知识整合论 (一). 医学信息 : 医学与计算机应用 , 16(6), 274-279.]

Bohr, N. (1964). Atomic theory and the description of nature (Yu, D. Trans.). Beijing: The Commercial Press. [N. 玻尔. (1964). 原子论和自然的描述 (郁韬 译). 北京 : 商务印书馆.]

Chang, J., & Ge, Y. (2005). Compendium for unified biology. Beijing: Higher Education Press. [常杰 , 葛滢. (2005). 统合生物学纲要. 北京 : 高等教育出版社.]

Chen, J., & Wu, Q. (2007). Study of industrial cluster by integration theory. Science & Technology Progress and Policy, 24(3), 58-61. [陈捷娜 , 吴秋明. (2007). 产业集群的集成论阐释. 科技进步与对策 , 24(3), 58-61.]

Chen, Y. (2008). The history and critical thinking of neural science. Shanghai: Shanghai Science and Technology Press. [陈宜张. (2008). 神经科学的历史发展和思考. 上海 : 上海科学技术出版社.]

Crick, F. (1994).What mad pursuit: A personal view of scientific discovery (Lü, X., & Tang, X. Trans.). Hefei: University of Science and Technology of China Press, 1994. [弗朗西斯·克里克. (1994). 狂热的追求 : 科学发现之我见 (吕向东 , 唐孝威 译). 合肥 : 中国科学技术大学出版社.]

Dai, R. (2009). Engineering innovation based on metasynthesis. Journal of Engineering Studies, 1(1), 46-50. [戴汝为. (2009). 基于综合集成法的工程创新. 工程研究 : 跨学科视野中的工程 , 1(1), 46-50.]

De Saussure, F. (1980). Course in general linguistics (Gao, M. Trans.). Beijing: The Commercial Press. [费尔迪南·德·索绪尔. (1980). 普通语言学教程 (高名凯 译). 北京 : 商务印书馆 .]

Eccles, J.C. (2004). Evolution of the brain: Creation of the self (Pan, H. Trans.). Shanghai: Shanghai Scientific and Technological Education Publishing House. [约翰·C. 埃克尔斯. (2004). 脑的进化 : 自我意识的创生 (潘泓 译). 上海 : 上海科技教育出版社 .]

Fang, J. (1982). Set Theory. Changchun: Jilin People's Publishing House. [方嘉琳. (1982). 集合论. 长春 : 吉林人民出版社 .]

Freeman, W.J. (2004). Neurodynamics: An exploration in mesoscopic brain (Gu, F., Liang, P. et al. Transs). Hangzhou: Zhejiang University Press. [沃尔特·J. 弗利曼. (2004). 神经动力学 : 对介观脑动力学的探索 (顾凡及，梁培基等 译). 杭州 : 浙江大学出版社 .]

Gintis, H., Bowles, S., et al. (2005). Towards a unified social science: A view from the Santa Fe school (ICSS, ZJU Trans.). Shanghai: Shanghai People's Publishing House. [赫伯特·金迪斯，萨缪·鲍尔斯等. (2005). 走向统一的社会科学 : 来自桑塔费学派的看法 (浙江大学跨学科社会科学研究中心 译). 上海 : 上海人民出版社 .]

Gu, F., & Liang, P. (2007). Neural information process. Beijing: Beijing University of Technology Press. [顾凡及，梁培基. (2007). 神经信息处理. 北京 : 北京工业大学出版社 .]

Hai, F., Li, B., & Feng, Y. (2001). Basic categories of integration theory. China Soft Science, 1, 114-117. [海峰，李必强，冯艳飞. (2001). 集成论的基本范畴. 中国软科学, 1, 114-117.]

Hai, F., & Li, B. (1999). Integration theory on management. China Soft Science, 3, 86-87. [海峰，李必强. (1999). 管理集成论. 中国软科学, 3, 86-87.]

Hu, Q. (2002). Integration theory on culture. Journal of Guizhou Ethnic Institute (Philosophy and Social Sciences), 1, 36-60, 53. [胡启勇. (2002). 文化整合论. 贵州民族学院学报 : 哲学社会科学版, 1, 36-40, 53.]

Huang, B. (2000). Brain's higher order function and neural network. Beijing: Science Press. [黄秉宪. (2000). 脑的高级功能与神经网络. 北京 : 科学出版社 .]

Lambright, W.H. (2009). The key of "big science": Managing and coordinating (Wang, X. Trans.). Beijing: Science Press. [Lambright, W.H. (2009). 重大科学计划实施

的关键：管理与协调 (王小宁 译). 北京：科学出版社 .]

Li, B., & Hu, H. (2004). Integration theory on enterprise property. Monthly Journal of Theory, 5, 165-166. [李必强，胡浩 . (2004). 企业产权集成论 . 理论月刊，5, 165-166.]

Liu, X. (1997). On the integration theory. China Soft Science, 10, 103-106. [刘晓强 . (1997). 集成论初探 . 中国软科学，10, 103-106.]

Lü, S. (1979). Analyses on Chinese grammar. Beijing: The Commercial Press. [吕叔湘 . (1979). 汉语语法分析问题 . 北京：商务印书馆 .]

Niu, S. (1997). On the integration of life: Exploring the phenomenon of life. Beijing: China Minzu University Press. [牛世盛 . (1997). 生命整合论：生命现象探索 . 北京：中央民族大学出版社 .]

Pan, S. (1985). Human intelligence: The human mind in pictures. Shanghai: Shanghai Science and Technology Press. [潘菽 . (1985). 人类的智能：人类心理图说 . 上海：上海科学技术出版社 .]

Peng, D. (2001). General psychology. Beijing: Beijing Normal University Press. [彭聃龄 . (2001). 普通心理学 . 北京：北京师范大学出版社 .]

Piaget, J. (2006). Structuralism (Ni, L., & Wang, L. Trans.). Beijing: The Commercial Press. [皮亚杰 . (2006). 结构主义 (倪连生，王琳 译). 北京：商务印书馆 .]

Qian, X. (1986). About scientific understanding. Shanghai: Shanghai People's Publishing House. [钱学森 . (1986). 关于思维科学 . 上海：上海人民出版社 .]

Qian, X. (2007). Create a systematics. Shanghai: Shanghai Jiaotong University Press. [钱学森 . (2007). 创建系统学 . 上海：上海交通大学出版社 .]

Qian, X., Yu, J., & Dai, R. (1990). A new discipline: Open complex giant system and methodology. Chinese Journal of Nature, 1, 4. [钱学森，于景元，戴汝为 . (1990). 一个科学新领域：开放的复杂巨系统及其方法论 . 自然杂志，1, 4.]

Song, X., Chen, F., & Tang, X. (2007). Unconsciousness and energy consumption of the brain in the resting state. Chinese Journal of Applied Psychology, 13(1), 33-36, 43. [宋晓兰，陈飞燕，唐孝威 . (2007). 无意识活动与静息态脑能量消耗 . 应用心理学，13(1), 33-36, 43.]

Song, X., & Tang, X. (2008). The extended global workspace theory of consciousness. Progress in Natural Science, 18(6), 622-627. [宋晓兰，唐孝威 . (2008). 意识全局工作空间的扩展理论 . 自然科学进展，18(6), 622-627.]

Tang, X. (1985). Electron-positron collider. Beijing: People's Education Press. [唐孝威 .

(1985). 正负电子对撞实验. 北京：人民教育出版社.]

Tang, X. (1992). On the force between chromosome during the telophase of mitosis. Progress of Nature Science, 2, 454-456. [唐孝威. (1992). 关于有丝分裂后期染色体作用力的讨论. 自然科学进展, 2, 454-456.]

Tang, X. (1993). L3 collaboration experiments. Progress of Nature Science, 3, 303-308. [唐孝威. (1993). L3合作实验. 自然科学进展, 3, 303-308.]

Tang, X. (1999). Functional brain imaging. Hefei: University of Science and Technology of China Press. [唐孝威. (1999). 脑功能成像. 合肥：中国科学技术大学出版社.]

Tang, X. (2001). Nuclear medicine and radiation therapy. Beijing: Beijing Medical University Press. [唐孝威. (2001). 核医学和放射治疗技术. 北京：北京医科大学出版社.]

Tang, X. (2003a). Principles of brain function. Hangzhou: Zhejing University Press. [唐孝威. (2003a). 脑功能原理. 杭州：浙江大学出版社.]

Tang, X. (2003b). Four-component theory of consciousness. Chinese Journal of Applied Psychology, 9(3), 10-13. [唐孝成. (2003b). 意识的四个要素理论. 应用心理学, 9(3), 10-13.]

Tang, X. (2004). On consciousness: Natural science research on consciousness. Beijing: Higher Education Press. [唐孝威. (2004). 意识论：意识问题的自然科学研究. 北京：高等教育出版社.]

Tang, X. (2005). Discussion on a grand unified theory of psychology. Chinese Journal of Applied Psychology, 11(3), 282-283. [唐孝威. (2005). 关于心理学统一理论的探讨. 应用心理学, 11(3), 282-283.]

Tang, X. (2007). Unified framework of psychology and cognitive science. Shanghai: Shanghai People's Publishing House. [唐孝威. (2007). 统一框架下的心理学与认知理论. 上海：上海人民出版社.]

Tang, X. (2008a). Unconscious mental activities. Hangzhou: Zhejiang University Press. [唐孝威. (2008a). 心智的无意识活动. 杭州：浙江大学出版社.]

Tang, X. (2008b). AMPLE intelligence model: Extension of PASS intelligence model. Chinese Journal of Applied Psychology, 14(1), 66-69. [唐孝威. (2008b). AMPLE智力模型：PASS智力模型的扩展. 应用心理学, 14(1), 66-69.]

Tang, X. (2010). On intelligence: Integration of mental and behavioral abilities. Hangzhou: Zhejiang University Press. [唐孝威. (2010). 智能论：心智能力和行

为能力的集成. 杭州 : 浙江大学出版社 .]

Tang, X., Du, J., Chen, X., Wei, E., Xu, Q., & Qin, L. (2006). An introduction to the brain. Hangzhou: Zhejiang University Press. [唐孝威 , 杜继曾 , 陈学群 , 魏尔清 , 徐琴美 , 秦莉娟 . (2006). 脑科学导论. 杭州 : 浙江大学出版社 .]

Tang, X., & Guo, A. (2000). Unified model of selective attention. ACTA Biophsica Sinica, 16(1), 187-188. [唐孝威 , 郭爱克 . (2000). 选择性注意的统一模型 . 生物物理学报 , 16(1), 187-188.]

Tang, X., & Huang, B. (2003). The theory of four functional systems of the brain. Chinese Journal of Applied Psychology, 9(2), 3-5. [唐孝威 , 黄秉宪 . (2003). 脑的四个功能系统学说. 应用心理学 , 9(2), 3-5.]

Tang, X., & Liu, G. (1992). Quantitative measurement of pollen tube growth and particle movement. Acta Botanic Sinica, 34(12), 893-898. [唐孝威 , 刘国琴 . (1992). 花粉管生长和内部颗粒运动的定量测量. 植物学报 , 34, 893-898.]

Tang, X., Shen, X., & He H. (2008). A hypothesis about the main and collateral channels of human body. Strategic Study of CAE, 10(11), 14-17. [唐孝威 , 沈小雷 , 何宏建 . (2008). 关于人体经络的一个试探性观点. 中国工程科学 , 10(11), 14-17.]

Tang, X., Wu, Y., Shan, B., & Zeng, H. (2001). Hypothesis of hierarchically associated codingbased on the neuronal clusters. Acta Biophysica Sinica, 17(4), 806-808. [唐孝威 , 吴义根 , 单保慈 , 曾海宁 . (2001). 神经元簇的层次性联合编码假设. 生物物理学报 , 17(4), 806-808.]

Von Bertalanffy, L. (1987). General system theory: Foundation, development, and applications (Qiu, T., & Yuan, J. Trans.). Beijing: Social Sciences Academic Press. [L. 贝塔朗菲. (1987). 一般系统论 : 基础·发展·应用 (秋同 , 袁嘉新 译). 北京 : 社会科学文献出版社 .]

Wei, Y., & Rowel, P. (2005). Guide for inquiry based science education. Beijing: Educational Science Publishing House. [韦钰 , P. Rowell. (2005). 探究式科学教育教学指导. 北京 : 教育科学出版社 .]

Wiener, N. (2007). Cybernetics: Or the control and communication in the animal and the machine (Hao, J. Trans.). Beijing: Peking University Press. [维纳 . (2007). 控制论 : 或关于在动物和机器中控制和通信的科学 (郝季仁 译). 北京 : 北京大学出版社 .]

Yan, L., Tang, X., & Liu, G. (1994). Flowing track system of cytoplasmic transportation in pollen tube. Progress in Natural Science, 4, 599-602. [阎隆飞 , 唐孝威 , 刘国琴 .

(1994). 花粉管胞质颗粒运输的流动轨道系统. 自然科学进展, 4, 599-602.]

Yang, N., Chen, Z., Lu, P., Zhang, C., Zhai, Z., & Tang, X. (2003). Assembly of stars in the oocyte acellular system of xenopus slaevis and its role in the nuclear membrane reconstruction. Science Bulletin, 48(15), 1623-1627. [杨宁, 陈忠才, 卢萍, 张传茂, 翟中和, 唐孝威. (2003). 非洲爪蟾卵非细胞体系中星体的组装及其在核膜重建中的作用. 科学通报, 48(15), 1623-1627.]

Yu, H., Li, H., & Lü, X. (2005). Integration theory of network organization. Monthly Journal of Theory, 2, 110-112. [喻红阳, 李海婴, 吕鑫. (2005). 网络组织集成论. 理论月刊, 2, 110-112.]

Zehna, P.W., & Johnson, R.L. (1986). Elements of set theory (Mai, Z., & Mai, S. Trans.). Beijing: Science Press. [P.W. 齐纳, R.L. 约翰逊. (1986). 集合论初步 (麦卓文, 麦绍文 译). 北京: 科学出版社.]

Zhang, X. (1997). Collected papers of Zhang Xiangtong: 1936—1997. Shanghai: The Library of Shanghai Branch, Chinese Academy of Sciences. [张香桐. (1997). 张香桐科学论文集: 1936—1997. 上海: 中国科学院上海分院图书馆.]

Foreign Language Reference

Adeva, B., Aguilar-Benitez, M., Akbari, H., Alcaraz, J., Aloisio, A., Alvarez-Taviel, J., ... Zoll, J. (1990). The construction of the L3 experiment. Nuclear Instruments and Methods in Physics Research Section A: Accelerators, Spectrometers, Detectors and Associated Equipment, 289(1-2), 35-102.

Alkire, M.T., Haier, R.J., Barker, S.I., & Shah, N.K. (1995). Cerebral metabolism during propofol anesthesia in humans studied with positron emission tomography. Anesthesiology, 82(2), 393-403.

Alkire, M.T., Haier, R.J., Shah, N.K., & Anderson, C.T. (1997). Positron emission tomography study of regional cerebral metabolism in humans during isoflurane anesthesia. Anesthesiology, 86(3), 549-557.

Alkire, M.T., Pomfrett, C., Haier, R.J., Gianzero, M.V., Chan, C.M., Jacobsen, B.P., & Fallon, J.H. (1999). Functional brain imaging during anesthesia in humans: Effects of halothane on global and regional cerebral glucose metabolism. Anesthesiology, 90(3), 701-709.

Anderson, J.R., Dan, B., Byrne, M.D., Douglass, S.A., Lebiere, C., & Qin, Y. (2004). An integrated theory of the mind. Psychological Review, 111(4), 1036-1060.

Anderson, J.R. (1983). The architecture of cognition. Massachusetts: Harvard University Press.

Arnold, M.B. (1960). Emotion and personality. New York: Columbia University Press.

Baars, B.J., & Franklin, S. (2003). How conscious experience and working memory interact. Trends in Cognitive Sciences, 7(4), 166-172.

Baars, B.J. (1983). Conscious contents provide the nervous system with coherent, global information. In Davidson, R.J., Schwartz, G.E., & Shapiro, D. Consciousness and self-regulation. New York: Plenum Press.

Baars, B.J. (1988). A cognitive theory of consciousness. New York: Cambridge University Press.

Barlow, H.B. (1972). Single units and sensation: A neuron doctrine for perceptual psychology? Perception, 1(4), 371-394.

Beggs, M.J., & Plenz, D. (2004). Neuronal avalanches are diverse and precise activity patterns that are stable for many hours in cortical slice cultures. The Journal of Neu-roscience, 24(22), 5216-5229.

Biswal, B.B., Mennes, M., Zuo, X.N., Gohel, S., Kelly, C., Smith, S.M., ... Milham, M.P. (2010). Toward discovery science of human brain function. Proceedings of the National Academy of Sciences, 107(10), 4734-4739.

Brooks, R.A. (1991). Intelligence without reason. In Mylopoulos, J., Reiter, R. Proceedings of the 12th International Joint Conference on Artificial Intelligence. San Mateo, CA: Morgan Kaufmann, 569-595.

Brooks, R.A. (1999). Cambrian intelligence: The early history of the new AI. Massachusetts: The MIT Press.

Buzsáki, G. (2006). Rhythms of the brain. New York Oxford University Press.

Buzsáki, G. (2007). The structure of consciousness. Nature, 446(7133), 267.

Cacioppo, J.T., Berntson, G.G., Adolphs, R., Carter, C.S., Davidson, R.J., McClintock M.K., ... Taylor, S.E. (2002). Foundations in social neuroscience. Massachusetts: The MIT Press.

Chomsky, N. (1957). Syntactic structure. The Hague: Mouton.

Corbetta, M., Miezin, F.M., Dobmeyer, S., Shulman, G.L., & Petersen, S.E. (1991). Selective and divided attention during visual discriminations of shape, color and speed: Functional anatomy by positron emission tomography. Journal of Neuroscience, 11, 2383-2402.

Crick, F., & Koch, C. (2003). A framework for consciousness. Nature Neuroscience, 6, 119-126.

Crick, F. (1984). Function of the thalamic reticular complex: The searchlight hypothesis. Proceedings of the National Academy of Sciences, 81(14), 4586-4590.

Das, J.P., Naglieri, J.A., & Kirby, J.R. (1994). Assessment of cognitive processes: The PASS theory of intelligence. Boston: Allyn and Bacon.

Dehaene, S. (2001). The cognitive neuroscience of consciousness. Massachusetts: The MIT Press.

Denmark, F.L., & Krauss, H.H. (2005). Unification through diversity. In Sternberg, R.J. Unity in psychology: Possibility or pipedream? Washington, DC: American Psychological Association.

Desimone, R., & Duncan, J. (1995). Neural mechanisms of selective visual attention.

Annual Review of Neuroscience, 18: 193-222.

Duncan, J., Ward, R., & Shapiro, K. (1994). Direct measurement of attentional dwell time in human vision. Nature, 369 (6478), 313-315.

Eckhorn, R., Bauer, R., Jordan, W., Brosch, M., & Reitboeck, H.J. (1988). Coherent oscillations: Amechanism for feature linking in the visual cortex. Biological Cybernetics, 60 (2), 121-130.

Edelman, G.M., & Tononi, G. (2000). A universe of consciousness: How matter becomes imagination. New York: Basic Books.

Flourens, P. (1960). Investigations of the properties and the functions of the various parts which compose the cerebral mass. In von Bonin, G. Some papers on the cerebral cortex. Springfield, IL.: Charles C. Thomas, 3-21.

Matthei, E.H., & Fodor, J. (1984). The modularity of mind: An essay on faculty psychology. Cambridge, MA: MIT Press.

Fox, M.D., Snyder, A.Z., Vincent, J.L., Corbetta, M., van Essen, D.C., & Raichle, M.E. (2005). The haman brain is intrinsically organized into dynamic, anticor-related functional networks. Proceedings of the National Academy of Sciences, 102(27), 9673-9678.

Frackowiak, R., Friston, K.J., Frith, C.D., Dolan, R.J., & Mazziotta J.C. (1997). Human brain function. San Diego: Academic Press.

Fransson, P. (2005). Spontaneous low-frequency BOLD signal fluctuations: An fMRI investigation of the resting-state default mode of brain function hypothesis. Human Brain Mapping, 26(1), 15-29.

Fujii, H., Ito H., Aihara, K., Ichinose, N., & Tsukada, M. (1996). Dynamical cell assembly hypothesis: Theoretical possibility of spatio-temporal coding in the cortex. Neural Network, 9(8), 1303-1350.

Fuster, J. (1997). Network memory. Trends in Neurosciences, 20(1), 451-459.

Gallagher, H., & Frith, C. (2003). Functional imaging of 'theory of mind'. Trends in Cognitive Sciences, 7(2), 77-83.

Gardner, H. (1985). The mind's new science: A history of the cognitive revolution. New York: Basic Books.

Gardner, H. (1993). Multiple intelligence: The theory in practice, A Reader. New York: Basic Books.

Gazzaniga, M.S., Ivry, R.B., & Mangun, G.R. (1998). Cognitive neuroscience: The biology of the life. New York: W. W. Norton & Company.

Gazzaniga, M.S. (2000). The cognitive neuroscience (2nd ed.). Cambridge: The MIT

Press.

Gibson, J.J. (1979). An ecological approach to visual perception. Boston: Houghton Mifflin.

Glassman, W.E. (2000). Approaches to psychology (3rd ed.). Philadelphia: Open University Press.

Goleman, D. (1995). Emotional Intelligence. New York: Bantam Books.

Gray, C.M., Knig, P., Engel, A.K., & Singer, W. (1989). Oscillatory responses in cat visual cortex exhibit inter-columnar synchronization which reflects global stimulus properties. Nature, 338, 334-337.

Greicius, M.D., Ben, K., Reiss, A.L., & Vinod, M. (2003). Functional connectivity in the resting brain: A network analysis of the default mode hypothesis. Proceedings of the National Academy of Sciences, 100(1), 253-258.

Greicius, M.D., & Menon, V. (2004). Default-mode activity during a passive sensory task: Uncoupled from deactivation but impacting activation. Journal of Cognitive Neuroscience, 16(9), 1484-1492.

Gusnard, D.A., & Raichle, M.E. (2001). Searching for a baseline: Functional imaging and the resting human brain. Nature Reviews Neuroscience, 2, 685-694.

Haberlandt, K. (1997). Cognitive Psychology (2nd ed.). Boston: Allyn and Bacon.

Hawkins, J.,& Blakeslee, S. (2004). On intelligence. New York: Henry Holt.

He, Y., Wang, J., Wang, L., Chen, Z.J., Yan, C., Yang, H., ... Zang, Y. (2009). Uncovering intrinsic modular organization of spontaneous brain activity in humans. PLoS ONE, 4(4), e5226.

Hebb, D. (1949). The organization of behavior: A neuropsychological theory. New York: John Wiley.

Hilgard, E.R. (1987). Psychology in America: A historical survey. New York: Harcourt Brace College Publishers.

Hillyard, S.A., & Picton, T.W. (1987). Electrophysiology of cognition. In Mountcastle, V.B., Plum, F., & Geiger, S.R. Handbook of Physiology: Vol 5. Baltimore: American Physiological Society.

Hothersall, D. (1984). History of psychology. Philadelphia: Temple University Press.

Jacob, F. (1977). Evolution and tinkering. Science, 196(4295), 1161-1166.

Kandel, E.R., Schwartz, J.H., & Jessell, T.M. (2000). Principles of neural science (4th ed.). New York: McGraw-Hill Medical.

Kosslyn, S.M., & Koenig, O. (1995). Wet mind: The new cognitive neuroscience. New York: The Free Press.

Kuhn, T. (1970). The structure of scientific revolution (2nd ed.). London: Cambridge University Press.

L3 Collaboration. (1993). Results from the L3 experiment at LEP. Physics Reports, 236(1-2), 1-146.

Lakoff, G., & Johnson, M. (1999). Philosophy in the flesh: The embodied mind and its challenge to Western thought. New York: Basic Books.

Lashley, K.S. (1929). Brain mechanisms and intelligence. Chicago: University of Chicago Press.

Laureys, S. (2005). The neural correlate of (un) awareness: Lessons from the vegetative state. Trends in Cognitive Science, 9(12), 556-559.

Lazarus, R.S. (1993). From psychological stress to the emotion: A history of changing outlooks. Annual Review of Psychology, 44, 1-22.

Le Doux, J. (1996). The emotional brain: The mysterious underpinning of emotional life. New York: Simon and Schuster.

Liljenström, H., & Århem, P. (2008). Consciousness transitions: Phylogenetic, ontogenetic, and physiological aspects. Amsterdam: Elsevier.

Logothetis, N.K., Pauls, J., Augath, M., Trinath, T., & Oeltermann, A. (2001). Neurophysiological investigation of the basis of the FMRI signal. Nature, 412(6843), 150-157.

Luria, A.R. (1966). Human brain and psychological processes. New York: Harper and Row.

Luria, A.R. (1973). The working brain: An introduction to neuropsychology. New York: Basic Books.

Maddock, R.J. (1999). The retrosplenial cortex and emotion: New insights from functional neuroimaging of the human brain. Trends in Neurosciences, 22(7), 310-316.

Maier, S.F., & Watkins, L.R. (1998). Cytokines for psychologists: Implications of bidirectional immune-to-brain communication for understanding behavior, mood, and cognition. Psychological Review, 105(1), 83-107.

McCann, S.M., Lipton, J.M., Sternberg, E.M., Chrousos, G.P., Gold, P.W., & Smith, C.C. (1998). Neuroimmunomodulation: Molecular aspects, integrative systems and clinical advances. Annals of the New York Academy of Sciences, vol.840.

McCarthy, R.A., & Warrington, E.K. (1990). Cognitive neuropsychology: A clinical introduction. London: Academic Press.

Mckiernan, K.A., Kaufman, J.N., Kucera-Thompson, J., & Binder, J.R. (2003).

A parametric manipulation of factors affecting task-induced deactivation in functional neuroimaging. Jourmal of Cognitive Neuroscience, 15(3), 394-408.

Melmed, S. (2001). Series Introduction: The immuno-neuroendocrine interface. Journal of Clinical Investigation, 108(11), 1563-1566.

Miall, R.C., & Robertson, E.M. (2006). Functional imaging: Is the resting brain resting? Current Biology, 116(23), 998-1000.

Miller, G. (2007). Hunting for meaning after midnight. Science, 315 (5817), 1360-1363.

Naccache, L. (2006). Is she conscious? Science, 313(5792), 1395-1396.

Naglieri, J.A., & Das, J.P. (1990). Planning, attention, simultaneous and successive (PASS) cognitive processes as a model for intelligence. Journal of Psychoeducational Assessment, 8(3), 303-337.

Newell, A., & Simon, H. (1972). Human problem solving. Englewood Cliffs, NJ: Prentice-Hall.

Newell, A. (1990). Unified theories of cognition. Massachusetts: Harvard University Press.

Northoff, G., Bermpohl F. (2004). Cortical midline structures and the self. Trends in Cognitive Sciences, 8(3), 102-107.

Piaget, J. (1983). Piaget's theory. In Mussem, P. Handbook of child psychology, vol.1(4th ed.). New York: Wiley.

Posner, M.I., & Rothbart, M.K. (2006). Educating the human brain. Washington DC: American Psychological Association.

Purves, D., Augustine, G.J., Fitzpatrick, D., Katz, L.C., LaMantia A.S., & McNamara, J.O. (1997). Neuroscience. Sunderland: Sinauer Associates.

Raichle, M.E., MacLeod, A.M., Snyder, A.Z., Powers, W.J., Gusnard, D.A., & Shulman, G.L. (2001). A default mode of brain function. Proceedings of the National Academy of Sciences, 98(2), 676-682.

Raichle, M.E., & Mintun, M.A. (2006). Brain work and brain imaging. Annual Review of Neuroscience, 29, 449-476.

Raichle, M.E., & Snyder, A.Z. (2007). A default mode of brain function: A brief history of an evolving idea. Neuroimage, 37(4), 1083-1090.

Raichle, M.E. (2006). The brain's dark energy. Science, 314(5803), 1249-1250.

Salovey, P., & Mayer, J.D. (1990). Emotional intelligence. Imagination, Cognition and Personality, 9(3), 185-211.

Schiff, N.D., Ribary, U., Moreno, D.R., Beattie, B., Kronberg, E., Blasberg, R., ... Plum, F. (2002). Residual cerebral activity and behavioural fragments can remain in the persistently vegetative brain. Brain, 125(6), 1210-1234.

Sdorow, L. (1995). Psychology (3rd ed.). Madison: Brown and Benchmark.

Searle, J. (2000). Consciousness. Annual Review of Neuroscience, 23, 557-578.

Shannon, C.E., & Weaver, W. (1949). The mathematical theory of communication. Urbana: Univesity of Illinois Press.

Sherrington, C.S. (1906). The integrative action of the nervous system. New York: Charles Scribner's Sons.

Shulman, G.L., Corbetta, M., Buckner, R.L., Fiez, J.A., Miezin, F.M., Raichle, M.E., ... Petersen, S.E. (1997). Common blood flow changes across visual tasks: Decreases in cerebral cortex. Journal of Cognitive Neuroscience, 9(5), 648-663.

Shulman, R.G., Hyder, F., & Rothman, D.L. (2004). Energetic basis of brain activity: Implications for neuroimaging. Trends in Neuroscience, 27(8), 489-495.

Singer, W., & Gray, C. (1995). Visual feature integration and the temporal correlation hypothesis. Annual Review of Neuroscience, 18, 555-586.

Staats, A.W. (1999). Unifying psychology requires new infrastructure, theory, method, and research agenda. Review of General Psychology, 3(1), 3-13.

Sternberg, R.J., & Grigorenko, E.L. (2001). Unified psychology. American Psychologist, 56(12), 1069-1079.

Sternberg, R.J. (1985). Beyond IQ: A triarchic theory of human intelligence. New York: Cambridge University Press.

Sternberg, R.J. (2005). Unity in psychology: Possibility or pipedream? Washington, DC: American Psychological Association.

Tang, X. (1998). Meshwork supported fluid film model of cell membranes. Chinese Physics Letter, 15(10), 770-771.

Thompson, E., & Varela, F.J. (2001). Radical embodiment: Neural dynamics and consciousness. Trends in Cognitive Science, 5(10), 418-425.

Tian, L., Jiang, T., Liu, Y., Yu, C., Wang, K., Zhou, Y., ... Li, K. (2007). The relationship within and between the extrinsic and intrinsic systems indicated by resting state correlational patterns of sensory cortices. Neuroimage, 36(3), 684-690.

Tononi, G., Sporns, O., & Edelman, G.M. (1994). A measure for brain complexity: Relating functional segregation and integration in the neurons system. Proceedings of the National Academy of Sciences, 91(11), 5033-5037.

Treisman, A., & Gelade, G. (1980). A feature-integration theory of attention. Cognitive Psychology, 12(1), 97-136.

Treisman, A., Sykes, M., & Gelade, G. (1977). Selective attention stimulus integration. In Dornie, S. Attention and performance VI. Hilldale NJ: Lawrence Erlbaum, 333-361.

Underleider, L., & Mishkin, M. (1982). Two cortical visual systems. In Ingle, D., Mansfield, R.J.W., & Goodale, M.S. Analysis of visual behavior. Cambridge MA: MIT Press, 549-586.

Varela, F.J., Thompson, E., Rosch, E., & Kabat-Zinn, J. (1991). The embodied mind: Cognitive science and human experience. Massachusetts: The MIT Press.

Vogeley, K., May, M., Ritzl, A., Falkai, P., Zilles, K., & Fink, G.R. (2004). Neural correlates of first-person perspective as one constituent of human self-consciousness. Journal of Cognitive Neuroscience, 16(5), 817-827.

Von Bertalanffy, L. (1950). An outline of general system theory. British Journal for the Philosophy of Science, 1(2): 139-165.

Von Bertalanffy, L. (1976). General system theory: Foundation, development, applications (Rev. ed.). New York: George Braziller.

Von der Malsburg, C. (1981). The correlation theory of brain function. Goettingen: Max-Planck-Institute for Biophysical Chemistry.

Waelti, P., Dickinson, A., & Schultz, W. (2001). Dopamine responses comply with basic assumptions of formal learning theory. Nature, 412, 43-48.

Wickelgren, I. (2003). Tapping the mind. Science, 299 (5606), 496-499.

Wilson, E. (1998). Consilience: The unity of knowledge. New York: Alfred A. Knopf.

Waelti, P., Dickinson, A., & Schultz, W. (2002). Brain-computer interfaces for communication and control. Clinical Neurophysiology, 113(6), 767-791.

Zeki, S. (2003). The disunity of consciousness. Trends in Cognitive Sciences, 7(5), 214-218.

Zeki, S. (1993). A vision of the brain. Oxford: Wiley-Blackwell.

Terminology

Accommodation: change the original structure to fit the new environment

Art integratics: one discipline studying the characteristics, principles, and applications of integration phenomenon in art

Assimilation: incorporate the new information into the original structure

Bio-integratics: one discipline studying the characteristics, principles, and applications of integration phenomenon in biological system

Brain environment integration: integration process of the brain and environment

Brain integratics: one discipline studying the characteristics, principles, and applications of integration phenomenon in brain

Brainics: one subdiscipline of bionics focusing on the brain

Cell integratics: one discipline studying the characteristics, principles, and applications of integration phenomenon in cell

Cogno-integratics: one discipline studying the characteristics, principles, and applications of integration phenomenon in cognitive process

Critical condition: the condition that quality change is induced by the change of quantity

Culture integratics: one discipline studying the characteristics, principles, and applications of integration phenomenon in cultural field

Economics integratics: one discipline studying the characteristics, principles, and applications of integration phenomenon in economics

Education integratics: one discipline studying the characteristics, principles, and applications of integration phenomenon in education

Emergence: new properties come out with the change of quality

Engineering integratics: one discipline studying the characteristics, principles and applications of integration phenomenon in engineering

Environment integratics: one discipline studying the characteristics, principles and applications of integration phenomenon in environmental science

General integratics: one discipline studying the general characteristics, principles, and applications of integration phenomenon

Geo-integratics: one discipline studying the characteristics, principles and applications of the integration phenomenon in geoscience

Grand unification: integrate all aspects of theory into one unity

Grand unified psychology theory: one theory unifying theory and hypothesis in every subdiscipline of psychology

Info-integratics: one discipline studying the characteristics, principles, and applications of integration phenomenon in information science

Integrated theory of cognition: cognitive theory from the view of integration

Integratics: one discipline studying the characteristics, principles, and applications of integration phenomenon

Intelligence integration: the integration of mental, behavioral, and specific capability

Intelligence integratics: one discipline studying the characteristics, principles, and applications of integration phenomenon in intelligent science

Knowledge integratics: one discipline studying the characteristics, principles, and applications of integration phenomenon in knowledge science

Knowledge integration: the integration process of knowledge

Med-integrates: one discipline studying the characteristics, principles, and applications of integration phenomenon in medical science

Mental interaction: the interaction between mental phenomenon

Modularity: assemble the standard module into a bigger module of the system

Module: standard building block of one system

Nervous integration: integration in nervous system

Neuro-integratics: one discipline studying the characteristics, principles, and applications of integration phenomenon in nervous system

Psycho-integratics: one discipline studying the characteristics, principles, and applications of integration phenomenon in psychological field

Psychological integration: integration of mental process

Reasonable reduction: divide the system into different interacting parts at a proper level and research the system at this level

Space integratics: one discipline studying the characteristics, principles, and applications of integration phenomenon in space science

Social integratics: one discipline studying the characteristics, principles, and applications of integration phenomenon in social fields

Technology integratics: one discipline studying the characteristics, principles, and applications of integration phenomenon in technological field

Unity: the holistic formed by integration

Epilogue

It was around 2014 when Professor Li Wu from the State Key Laboratory of Cognitive Neuroscience and Learning at Beijing Normal University asked me if I could translate Professor Tang Xiaowei's book *General Integration Theory* into English. Owing to the extensive topics researched by Professor Tang and the wide range of knowledge the book covers from multiple disciplines, it was not easy to find a suitable translator. Knowing that I am involved in systems science research with a multidisciplinary background, Professor Li asked me if I would like to do the job. I knew that Professor Tang is a renowned physicist who has made significant contributions to China's nuclear science and engineering, and has made important contributions in the interdisciplinary fields of biology, medicine, neuroscience, psychology, etc. Due to the limitations of my intellectual background and English proficiency, I was worried that I would not be qualified to translate Professor Tang's book, so I suggested having a try at first. Professor Tang quickly sent me a sample copy of *General Integration Theory*. I was deeply attracted and impressed by the content and viewpoints of the book, and decided to overcome the difficulties and take on this daunting translation task.

Based on in-depth research on the human brain, the most complex system in nature, General Integration Theory proposes the important concept of "integration" in complex systems, which is an optimization process that reflects the process of constructing a unified body of coordinated activities with new functions based on the interactions of a large number of integrated components in complex systems. The key point of this concept is the construction of a unified whole with new functions of lower-level structural units, functional units, and informational units, to construct macro-functions of higher levels or even the whole complex system through interaction. The integration of the human brain includes the structural and functional integration, informational integration, and psychological integration. From a developmental perspective, the human brain has the integration processes of nervous system differentiation, brain segmentation, neuronal migration, synapse formation, and circuit construction. Structurally, the human brain ranges from the microscopic level of molecules, subcellular units, cells, to the mesoscopic level of groups of neurons and local neural circuits, and to the macroscopic level of functional regions, functional systems and the whole brain structure, involving the integration at multiple spatial scales from less than a micron to a decimeter, and from micro to macro levels into a unified whole. The integration of brain structures provides a basis for the integration of functions, and the integration of functions, in turn, changes the integration of structures, such as the plasticity of structures from synapses to brain areas. Numerous neural circuits in the brain form functional subsystems with different functions that are relatively independent yet interconnected. Many functional subsystems are widely connected in different ways to form the overall brain network, including the four functional systems that ensure, regulate tension and arousal states, perceive, process and store information, formulate programs, regulate and control psychological activities and behaviors, and evaluate information and generate emotional experiences as well through overall synchronization and information transmission at the macro level. The integration of brain structure and function is accompanied by the integration of information, from a single neuron integrating input signals into action potentials, to hierarchical joint encoding of information by clusters of neurons, thereby expressing, processing, and even generating new information, and/or new concepts. The integrative function and process of mental activities with interactive components such as arousal attention, cognition, emotion and will, and self-consciousness are referred to as macro-level psychological integration.

General Integration Theory starts from the integration of the brain, examines the integration phenomena in nature, technology, and socioeconomic fields, and points out that integration phenomena are common in complex systems. In the integration process of complex systems, many integrated components organize into a unified whole of coordinated activities through their interactions with each other and the environment in certain contexts. General Integration Theory systematically proposes to construct modules and networks of different levels and properties according to global goals, leading to integrated products, reducing complex parts in a meaningful way, and then organically integrating them based on inherent connections and actions. It elaborates on the relationships of global and module, reduction and synthesis, binding and union, reconstruction and optimization, threshold and emergence, complementation and coordination, compliance and synchronization, and adaptation and assimilation in the integration process of complex systems.

General Integration Theory attempts to understand and construct how complex systems form a unified whole with coordinated functions through the process of integration. In methodological terms, it is totally different from the reductionist method that has achieved brilliant success in scientific research, attempting to transcend reductionism in the research of complex systems. Renowned physicist Anderson pointed out that the fundamental difficulty in the study of complex systems is "The ability to reduce everything to simple fundamental laws does not imply the ability to start from those laws and reconstruct the universe." Even if we separate or isolate each component from the system for a separate study, we still cannot reconstruct the phenomena and functions that emerge at the system level. "The constructionist hypothesis breaks down when confronted with the twin difficulty of scale and complexity. [...] at each level of complexity, entirely new properties appear, and the understanding of the new behaviors requires research which I think is as fundamental in its nature as any other. [...] At each stage, entirely new laws, concepts, and generalizations are necessary, requiring inspiration and creativity to just as great a degree as the previous one. Psychology is not applied biology, nor biology is applied chemistry." [Anderson, P. W. (1972). More is different: Broken symmetry and the nature of the hierarchical structure of science. *Science, 177*(4047), 393-396.] General Integration Theory responds directly to the problem raised by Anderson, pointing out that the integrated components of complex systems form a coordinated whole through interaction, and complex systems generate their properties and laws at different levels.

A key issue in General Integration Theory is the mechanism by which lower-level components in complex systems integrate into higher-level components and form new phenomena and laws through interaction. Many other disciplines have long been concerned with this question. For example, in statistical physics, when the order parameter reaches a critical value, long-range correlation is formed in the system, and fluctuations drive the system to undergo a critical phase transition. The macroscopic properties formed by the interaction between the components of the system are completely different before and after the phase transition. In open non-equilibrium dissipative systems, the nonlinear interaction among the components of the system leads to the formation of specific spatiotemporal structures at the macro level, which are so-called dissipative structures. The core scientific problem of systems science research is concerned with the relationship between structure and function in complex systems, the rules of system evolution and its regulation, especially the mechanism of the formation of structure and function at the macro-level through the interaction among microscopic components in complex systems. It can be said that integration theory and systems science are similar.

The General Integration Theory provides a method for observing and thinking about complex systems that goes beyond reductionism. Of course, this method itself still needs further developing. In particular, we should know how to implement the thought process of integration theory in a quantitative and operational way to specific complex systems. Just as Prigogine used nonlinear dynamical equations to characterize chemical reactions in solutions and showed that the Belousov-Zhabotinsky reaction produces dissipative structures with spatiotemporal oscillations through bifurcation behavior, integration theory also needs to use modern mathematical tools or even develop new analytical tools in order to show the integration process of the coordinated activities of the unity of the whole constructed through interactions between integrated components in complex systems. Taking brain integration as an example, one possible method is to demonstrate its integration process by rebuilding the human brain. The processes that take place in the brain are physical and chemical processes. These processes can be characterized by differential equations. For example, the membrane potential of neurons can be characterized by Hodgkin-Huxley equations, and the synapses mediating neuronal interactions can also be characterized by differential equations. Even the action of neurotransmitters can be characterized by equations. From a mathematical point of view, the brain is an ultra-high-dimensional nonlinear dynamical system. Therefore, the integration process and the integration action of the brain can be demonstrated by studying this ultra-high-dimensional dynamical system, e.g., the limit set of the system, the behavior of the system on the limit set, the relationship between the limit set of the system and the structure of the system, the change of the limit set of the system with external stimuli, and so on. Obviously, this is not an easy task, but with the rapid accumulation of data on neural connections in the brain and the rapid increase in high-performance computing capabilities, the process of digitizing the human brain using equations will accelerate, causing breakthroughs in the near future. At that time, we will have a clearer and more intuitive understanding of the brain's integration processes and integration actions. Similarly, suitable mathematical tools need to be worked out for us to find out the integration processes and actions of other complex systems.

Thanks to my courage, I finally finished translating Professor Tang's book. I am grateful that I had the opportunity to study the General Integration Theory in the course of my translation, and then understand the profound meaning and important value of the General Integration Theory. It gave me a completely new perspective and shed light on the study of complex systems.

In order to complete this translation work as soon as possible, I asked Tan Bo, a master's student majoring in English, and Zhang Mengya, my doctoral student, to help me with the translation. During the translation process, Tan Bo and I translated by chapters and paragraphs, and read the translations and the original text back to back, and then Zhang Mengya read and checked the full text of the translation. After several rounds of revision, the first draft of the translation was produced. Lastly, Professor Wang Xiaolu at the School of International Studies of Zhejiang University, and now a distinguished professor at the School of Foreign Languages of Hangzhou City University revised and polished the translation.

Given my limited knowledge, there are bound to be mistakes and omissions in the translation. Please do not hesitate to criticize and advise on any shortcomings.

WANG Dahui
DEC 2023

© **Zhejiang Education Publishing House 2023**

This publication is in copyright. Subject to statutory exception and to the provisions of relevant collective licensing agreements, no reproduction of any part may take place without the written permission of Zhejiang Education Publishing House.

First published 2023

ISBN 978-7-5722-6835-9

Editor in charge: WANG Huijie, JIANG Lei
Proofreader in charge: ZHU Chenhang
Art editor: HAN Bo
Technical editor: SHEN Jiuling
Book design: GU Ye @ Rongxiang Studio
Typeset by: Hangzhou Linzhi Advertising Ltd
Printed in China at Hangzhou JPP Ltd